Lewis Wright

Light, a Course of experimental Optics Chiefly with the Lantern

Lewis Wright

Light, a Course of experimental Optics Chiefly with the Lantern

ISBN/EAN: 9783337254766

Printed in Europe, USA, Canada, Australia, Japan

Cover: Foto ©berggeist007 / pixelio.de

More available books at **www.hansebooks.com**

LIGHT

A COURSE OF

EXPERIMENTAL OPTICS

CHIEFLY WITH THE LANTERN

BY

LEWIS WRIGHT

WITH ILLUSTRATIONS

London
MACMILLAN AND CO.
1882

The Right of Translation and Reproduction is Reserved

"ASSUREDLY there is something in the phenomena of Light; in its universality; in the high office it performs in Creation; in the very hypotheses which have been advanced as to its nature; which powerfully suggests the idea of the *fundamental*, the *primeval*, the antecedent in point of rank and conception to all other products or results of creative power in the physical world. It is LIGHT, and the free communication of it from the remotest regions of the Universe, which alone can give, and does fully give us, the assurance of a uniform and all-pervading Energy—a MECHANISM almost beyond conception complex, minute, and powerful, by which that influence, or rather that movement, is propagated. Our evidence of the existence of gravitation fails us beyond the region of the double stars, or leaves us at best only a presumption amounting to moral conviction in its favour. But the argument for a unity of design and action afforded by Light, stands unweakened by distance, and is co-extensive with the Universe itself."—SIR JOHN HERSCHEL.

PREFACE.

VERY recently, at the Annual Meeting (1882) of the Teachers' Training and Education Society, **Prof.** Huxley reiterated, in his capacity of an Examiner in the Science and Art Department, a complaint well known **at** South Kensington as to the great difference between two kinds of "what persons called knowledge." He remarked for the hundredth time, how "That which they had constantly **to** contend against in the teaching of science in this country was, that many teachers had no conception of that distinction; for they thought it quite sufficient to be able to repeat a number of scientific propositions and to get their pupils **to** repeat them as accurately as they themselves did. . . . . The teacher should be instructed that his business in teaching was to convey clear and vivid impressions of the body of facts upon which the conclusions drawn from those facts were based." To do this in some degree for one branch of Physics is the sole object of this unpretending little work; which hopes to be rather a handbook, companion, and supplement to the excellent **existing text-books,** than anything else. With perhaps one **exception in a** short explanation **of the** law of sines, no **set formulæ** whatever will be found **in it;** but **I have** merely tried to place clearly **before the mind of the reader,** though something like a complete **course of** actual experiments, the *physical realities* which underlie the phenomena **of Light** and Colour. **As helps,** there **are solely employed simple** mechanical analogies and a **few** diagrams, explained in language which it is hoped may be found in **reality simple and clear,** though not intended **to be** childish, or to debar any private student from the healthful exercise of **now and then considering** what the writer means.

Now in carrying **out a course of** experiments in Physical Optics, projection **upon a** screen **is** not only far superior in general effect to any other method of demonstration, besides having the advantage of exhibiting the phenomena to the whole of **a class or audience** at the same time, but has another recommendation of primary importance. The trained physicist well understands **the meaning of** the rays **which** enter his eye, **and are visible to him** solely, direct **from his** lens, prism, or **other apparatus;** but the scholar **new to the** subject finds it very difficult to interpret in terms **of physical phenomena** what thus appeals merely to his own **visual** impressions. **When he** looks through a prism, for instance, it **is** difficult **for him to** get rid of the

PREFACE.

vague notion that "something in the prism" colours **what he sees.** But when actual rays of light are projected through the prism, and the colours appear on the screen *apart from himself*, as it were, he cannot help understanding that what thus appears to others at the same moment **as to him is a** physical reality which he has to trace out, and which he must reduce to physical terms. **What is meant will be** readily understood by any science-teacher.

Hence the method of experiment **chiefly** adopted here, for which all science-teachers **and students** are deeply indebted to Professor Tyndall, **who carried lantern** demonstration to an extent and perfection never before attained. The magnificent apparatus of the **Royal Institution, however,** appears to have created an impression that electric **cameras** and other very costly appliances are necessary for effective work of this class; whereas the greater number of experiments can be shown satisfactorily to at least a science class with even a good gas-burner; while any good lantern can be made at small expense a very efficient piece of apparatus. To make this also clear, and thus induce many **teachers to** substitute the most perfect kind of demonstration for far less striking methods, or for mere diagrams, is a second object of these pages. At the same time I have tried not to forget the private student, and not only to give sufficient hints for him to make most of the experiments without lantern or other bulky apparatus, but especially to find abundant manual *work* for him, more particularly in the fascinating domain of polariscope phenomena.

I cannot but hope room may be found for such a work as this, when I recall how much my own delight in the experimental study of Optics has been increased by the individual help and teaching of a few friends. It is never easy, and but seldom possible, **to** acknowledge **all such** obligations; but mine have been very special in two directions. To the Rev. Philip **R.** Sleeman, F.R.A.S., **F.R.M.S., I** was not **only indebted before** this little **book was thought of,** but have been since, for many references to foreign papers and memoirs, and other "detached" information which only his wide and general acquaintance with the whole range **of** Physical Optics could have supplied. And to **Mr. C. J. Fox,** F.R.M.S., I owe my entire practical education in that mica-film work, which I hope others may find as attractive as I have done, and which so admirably illustrates the phenomena of polarised light. But for these two friends, some of what are likely **to be the most** acceptable among **the** following pages would never have been written.

In regard to the experiments described, there are two **things to be** said. It would have been desirable, if possible, to **have stated** the originator of every experiment; but it was not **possible.** Attempt has been made to **indicate,** as far as known, the first to employ any striking very *recent* experiment: but many of great beauty seem now such common property, that it is difficult to ascertain **who** first made them, or first adapted them for projection. I strongly suspect that we owe to Professor Tyndall many more than it has been possible categorically to ascribe **to him**; and **am** the more anxious **to** state this, because **his just** claims **in** higher matters appear to me almost studiously ignored by certain Continental physicists. Some arrangements are, to the best of my belief, original; but none are put forth as such except one or two expressly stated, and **it** should be perfectly understood that no personal claim is implied regarding any other experiments because no credit **is** given to some one else: the absence of such credit is simply due to sheer ignorance and the difficulty of acquiring knowledge concerning such matters.

The other remark is, that the **order of** the experiments differs considerably in some cases from that usually adopted. All that can be said on that point is, that such **is** the result of considerable reflection, and in the belief that the order chosen is, upon the whole, best adapted to the primary end of assisting vivid conception of the physical realities considered, and *the relation of the phenomena to one another*. Also, while **no** attempt is made to arrange the experiments in set "Lectures," the order followed is believed to lend itself best to such **a** connected course of experimental lectures as a teacher would desire to give to his class, extended or abridged as the case may require. **I** am not without hope that in such an extended course of experiments as are here collected for his choice, some hard-worked **teacher** may find real help in this respect. The same may be said as to the brief references made **to** the connection between the phenomena of Light and the problems of Molecular Physics. Brief as they are, it is hoped they may in some minds excite a real interest in those problems, and deepen that sense of the *reality* of the phenomena **which is** so desirable.

It is right to say, that a large number **of** the experiments **here** described appeared originally as a series of articles in *The English Mechanic*. The cordial welcome of these articles, **and requests** from all quarters couched in unusually flattering terms, did in fact lead to their extension —**for the additions are very** large indeed—into the present **work** ; and are chiefly responsible for its being attempted by one who has no claim to be considered anything but an **amateur, chiefly** desirous of showing others what can be accomplished with only small means. I am indebted to the proprietors **of the** journal above named for their free permission not only to use **the text of** the articles in **any way** thought best, but also the diagrams by which they **were illustrated.**

Some may think that apology **might be** offered for the concluding **chapter. None such will be, or** ought to be offered. **The** irresistible propensity to **go beneath the** surface and search **for the** hidden **essence** of things, has

PREFACE.

been felt and manifested not only by all our leading physicists, but by all who have had any vivid impressions concerning the mysteries surrounding them; and "a man's thoughts are as children born to him, which he may not carelessly let die." No one is bound to accept my thoughts; and since there is no danger that my name will give any factitious weight to them, I have a right to utter them as well as another, provided it be done without dogmatism or offence. If neither of these can be charged to me, the rest may take care of itself; and I shall only frankly confess that it is largely such thoughts as these—thoughts which have led me to regard Light as a possible Revealer of much more than many think—which have for years made its phenomena, to me, such a profoundly interesting study.

LEWIS WRIGHT.

LONDON, *May* 1, 1882.

CONTENTS.

CHAPTER I.
THE LANTERN AND ACCESSORY APPARATUS.

PAGE

The Lantern—The Optical Objective—Gas Burners—Mineral Oil Burners—The Lime-light—Advantages of "Regulators"—Centering the Light—Mounting the Lantern—Focusing Lens—Prism—Plane Reflector—Other Apparatus—Screens—Diagrams—Vertical Lantern. . . . . 1—21

CHAPTER II.
RAYS AND IMAGES.—REFLECTION.

Rays of Light—Rays form Images—Inversion of Images—Shadows—Law of Intensity at various distances—Law of Reflection—Virtual and Multiple Images—The Doubled Angle of Reflection—Application of this in the Reflecting Mirror—Reflection from Concave and Convex Mirrors—Images produced by Concave Mirrors—Scattered Reflection—Light Invisible . . . . . . 22—48

CHAPTER III.
REFRACTION.—TOTAL REFLECTION.—PRISMS AND LENSES.

The Refraction or Bending of Rays—The Law of Sines—Index of Refraction—Total Reflection—The Luminous Cascade—Prisms—Lenses—Images produced by Lenses—Focus of a Lens—Virtual Images and Foci . . . . 49—64

CHAPTER IV.

DISPERSION AND THE SPECTRUM.—DIFFERENT COLOURS HAVE DIFFERENT REFRANGIBILITY.

The Spectrum—Different Colours differently Refracted—and each Colour has its own Angle of Total Reflection—Position of the Prism and its Effect—Correction of Aberrations by Variations in Position—White Light a Compound of Various Colours—Suppression of Colour produces Colour—Artificial Composition of White Light—A Narrow Slit necessary for a Pure Spectrum—The Rainbow—Refraction and Dispersion not Proportional—Achromatic Prisms and Lenses—Direct Vision Prisms—Anomalous Dispersion . . . 65–87

CHAPTER V.

WHAT IS LIGHT?—VELOCITY OF LIGHT.—THE UNDULATORY THEORY.

Light has a Velocity—Velocity implies Motion of some sort—The Emission Theory—Transmission or Motion of a State of Things—Transmission of Wave Motion — Illustrations—Wave Motion and Vision—Analysis of Wave Propagation—The Ether—Refraction according to the Wave Theory—Total Reflection, Dispersion, and Anomalous Dispersion—Mechanical Illustrations 88–110

CHAPTER VI.

COLOUR.

Absorption of Colours—What it means—Absorbed, Reflected, and Transmitted Colours—Complementary Colours—The Eye cannot judge of Colour-waves—Mixtures of Lights and Pigments, and their Difference—Primary Colour Sensations, not the same thing as Primary Colours—Colour as we see it only a Sensation—Experiments showing merely Sensational Colour 111–128

CHAPTER VII.

SPECTRUM ANALYSIS.

Absorption Spectra—Their use in Analysis—Continuous Spectra—The Solar Spectrum—Line Spectra—Reversed Lines—Radiation and Absorption Reciprocal—Fraunhofer's Lines—Reversed Solar Lines—Thickened Lines—Solar, Stellar, and Planetary Chemistry . . . . . 129—139

CHAPTER VIII.

PHOSPHORESCENCE.—FLUORESCENCE.—CALORESCENCE.

Effects of Absorbed Vibrations—The Invisible Rays of the Spectrum—Three Independent Spectra non-existent—Phosphorescence—Fluorescence—Calorescence—Relation of Fluorescence to Phosphorescence . . . . 140—153

CHAPTER IX.

INTERFERENCE.

Net Result of Two Different Forces—Liquid and Tidal Waves—Why Single Interferences are not Traceable in Light—Interference of Sound Waves—Thin Films of Turpentine, Transparent Oxide, Soap, Water, and Air—Colour Dependent on Thickness of the Film—Proved to be Dependent also on Reflection from both Surfaces—Spectrum Analysis of Films—Soap Films and Sound Vibrations—Fresnel's Mirrors—Fresnel's Prism—Irregular Refraction—Diffraction—Gratings—Telescopic Effects—Other Simple Experiments in Diffraction—Striated Surfaces—Barton's Buttons—Nature of Interference Colours—Measurement of Waves—The Size of Molecules of Matter . . 154—199

APPENDIX TO CHAPTER IX.

Diffraction in the Microscope—Note on the Colours of Thick Plates . . . . 200—207

CHAPTER X.

DOUBLE REFRACTION AND POLARISATION

Double Refraction—Huyghens' Experiment of Reduplicating Images—Polarisation—Polarisers, Analysers, and Polariscopes—Phenomena of Tourmalines—Polarisation by Reflection and Refraction — What Polarisation implies — Analysis of Polarisation by Reflection and Refraction—The Polarising Angle—Analysis of Polarisation by Double Refraction—Extraordinary and Ordinary Wave-shells in Double Refracting Crystals—Action of the Tourmaline . 208—233

APPENDIX TO CHAPTER X.

The Vibrations of Common Light . . . . 234—241

CHAPTER XI.

POLARISING APPARATUS.

The Nicol Prism—Prazmowski's Modification—Large and Small Analysers—Nicol Prism Polariscope—Foucault's Prism—Care of Prisms—Glass Piles—The ordinary Lantern Polariscope—Simple Apparatus for Private Study—Norremberg's Doubler—Note on Artificial Nicol Prisms . . 242—257

CHAPTER XII.

CHROMATIC PHENOMENA OF PLANE-POLARISED LIGHT.—LIGHT AS AN ANALYSER OF MOLECULAR CONDITION.

Resolution of Vibrations—Interference Colours—Why Opposite Positions of the Analyser give Complementary Colours—Coloured Designs in Mica and Selenite—Demonstrations of Interference—Crystallisations—Organic Films — Effects of Strain or Tension—Effects of Heat—and of Sonorous Vibration . . . . 258—282

CHAPTER XIII.

ROTATORY POLARISATION.

Phenomena of Quartz—Right and Left-handed Quartz—Circular Waves—Effect of Retardation in one such Wave—Rotation of the Plane of Polarisation—Use of a Bi-quartz—Rotation in Fluids—The Saccharometer—Chromatic Effects of Quartz—Other Rotatory Crystals—Electro-Magnetic Rotation—Important Difference between Magnetic and other Rotations—Connection between Rotation and Molecular Constitution
                                                    283—293

CHAPTER XIV.

CIRCULAR AND ELLIPTIC POLARISATION.

Fresnel's Rhomb—Composition of two Rectilinear Vibrations into a Circular one—Quarter-Wave Plates—Other Methods of Producing Circular Polarisation—Rotational Colours of Circularly-Polarised Light—Reusch's Artificial Quartzes—Behaviour of Quartz in Circular Light—Phenomena of Thin Films when Analysed as well as Polarised Circularly . 294—307

CHAPTER XV.

OPTICAL PHENOMENA OF CRYSTALS IN PLANE-POLARISED LIGHT.

Rings in Uni-axial Crystals—Cause of the Black Cross—Apparatus for Projection or Observation—Preparation of Crystals—Artificial Crystals—Anomalous Dispersion in Apophyllite Rings—Quartz—Bi-axial Crystals—Apparatus for Wide-angled Bi-axials—Anomalous Dispersion in Bi-axials—Fresnel's Theory of Bi-axials—Deductions from it—Mitscherlich's Experiment—Conical and Cylindrical Refraction—Relations of the Axes in Uni-axials and Bi-axials—Composite, Irregular, and Hemitrope Crystals—Mica and Selenite Combinations—Crossed Crystals—Norremberg's Uni-axial Mica Combinations—Airy's Spirals—Savart's Bands . . . . 308—332

CHAPTER XVI.

OPTICAL PHENOMENA OF CRYSTALS IN CIRCULARLY POLARISED LIGHT.

Modifications in Crystal Figures Produced by one Quarter-Wave Plate—Explanation of the Phenomena—Results of Polarising and Analysing Circularly—Quartz in Circularly-Polarised Light—Spiral Figures Showing the Relation of Uni-axial and Bi-axial Axes . . . . . . 333—343

CHAPTER XVII.

POLARISATION AND COLOUR OF THE SKY.—POLARISATION BY SMALL PARTICLES.

Polarisation of the Sky—Light Polarised by all Small Particles—Blue Colour similarly Caused—Polarisation by Black Surfaces—Experimental Demonstration of the Phenomena—Multi-coloured Quartz Images—Identity of Heat, Light, and Actinism . . . . 344—351

CHAPTER XVIII.

CONCLUSION . . . . 352—358

INDEX 361

LIST OF ILLUSTRATIONS.

FIG.		PAGE
1.	Optical Objective	3
2.	Woolf Top for Wash Bottle	8
3.	Blow-through Jet	9
4.	U-tube Test	10
5.	Revolving Tripod Stand	12
6.	Socket	13
7.	Focusing Lens	14
8.	Reflector	15
9.	Glass Prism	15
10.	Socket	16
11.	Adjustable Table-stand	16
12.	Bunsen Holder	17
13.	Vertical Attachment	20
14.	Image formed by a Small Hole	23
15.	Inversion of Image	25
16.	Shadows and Law of Squares	26
17.	Reflection	27
18.	Reflected Image	29
19.	Multiple Images	30
20.	Symmetrical Multiple Images	31
21.	Kaleidoscope	32
22.	Pulse made Visible	33
23.	Motions produced by Heat made Visible	34
24.	Lissajous' Experiment	35
25.	The Kaleidophone	36
26.	Sound-ripples in a Glass	38
27.	Dolbear's Opheidoscope	39
28.	Concave Mirror	40
29.	Image formed by Concave Mirror	41
30.	Inverted Image	42

LIST OF ILLUSTRATIONS.

FIG.		PAGE
31.	Reflection from Unpolished Surfaces	46
32.	Scattered Reflection	48
33.	Refraction Tank	50
34.	Refraction and Reflection	50
35.	The Law of Sines	52
36.	Refraction into a Rarer Medium	53
37.	Total Reflection	54
38.	Total Reflection	54
39.	Two-necked Glass Receiver	55
40.	Luminous Cascade	56
41.	Deflection	57
42.	Deflected Image	58
43.	Prism	59
44.	Nature of a Lens	60
45.	Lens and Focus	61
46.	Image formed by Lens	62
47.	Lens and Virtual Image	63
48.	Concave Lens	63
49.	Production of the Spectrum	66
50.	Newton's Experiment	68
51.	Effects of Position in Lenses	71
52.	Recomposition of White Light	72
53.	Small Mirror	73
54.	The Colours Recompounded	73
55.	Newton's Colour Disc	74
56.	Blackened Disc	74
57.	Rocking Spectrum	77
58.	Rainbow Experiment	79
59.	Diagram of the Rainbow	81
60.	Water Prism	82
61.	Ahrens' Direct Bi-sulphide Prism	84
62.	Bi-sulphide Compound Prism	84
63.	Fizeau's Experiment	89
64.	Light necessarily Motion	91
65.	Trough for Ivory Balls	94
66.	Slide for Rolling Balls	95
67.	Centres for Crova's Disc	97
68.	Crova's Disc	97
69.	Wave-line Disc	98
70.	Wave Slide	98

LIST OF ILLUSTRATIONS.

FIG.		PAGE
71.	Movable part of Wave Slide	99
72.	Sound Waves and Light Waves	100
73.	Fallacy of Radial Wave Theory	101
74.	New Centres of Wave Motion	101
75.	True Nature of Wave Propagation	102
76.	Refraction	104
77.	Vibrations in a Glass	106
78.	Pair of Tylor's Wheels	108
79.	Mechanical Illustration of Refraction	109
80.	Mechanical Illustration of Dispersion	109
81.	Mechanical Illustration of Total Reflection	110
82.	Spectrum Work	112
83.	Slits for Complementary Colours	117
84.	Complementary Absorptions	120
85.	Subjective Colours	125
86.	(1) Iodine Vapour. (2) Nitrous Gas	130
87.	Reversed Sodium Line	134
88.	Solar F Line Reversed	137
89.	Fluorescence	146
90.	Lantern Phosphoroscope	152
91.	Interference of Liquid Waves	155
92.	Two Lighthouses	157
93.	Reflections from a Film	159
94.	Film of Turpentine	160
95.	Film of Oxide	160
96.	Rings for Soap Solution	163
97.	Flat Soap Film	165
98.	Newton's Lenses	166
99.	Newton's Rings with Flat Glasses	166
100.	Newton's Rings	167
101.	Analysis of the Rings	171
102.	Spectrum Analysis of the Rings	171
103.	Lantern Phoneidoscope	175
104.	Phoneidoscope Effects	177
105.	Fresnel's Mirrors	179
106.	Parallel Slits	180
107.	Fresnel's Prism	180
108.	Tablet for Objects	189
109.	Nature of Diffraction	192
110.	Measurement of Wave-length	194

FIG.		PAGE
111.	Dense and Rare Media	195
112.	Microscopic Diffraction Spectra	201
113.	Microscopic Diffraction Spectra	201
114.	Microscopic Diffraction Spectra	201
115.	Stopping out Spectra	202
116.	Obliteration of Structure	202
117.	Partially Stopping Spectra	202
118.	Apparent Creation of Structure	202
119.	Further Stopping of Spectra	203
120.	Apparent Multiplication of Structure	203
121.	Crossed Grating	203
122.	Rectangular Spectra	203
123.	Selection from Crossed Spectra	204
124.	Apparent Rotation of Structure	204
125.	Spectra of *Pleurosigma Angulatum*	204
126.	Corresponding Structure	204
127.	Calculated Image of *P. Angulatum*	205
128.	*Amphipleura Pellucida*	207
129.	Iceland Spar or Calcite	209
130.	Double Refraction	209
131.	Huyghens' Apparatus	210
132.	Huyghens' Experiment	210
133.	Analysis of Huyghens' Experiment	212
134.	Two Tourmalines	214
135.	Tourmaline and Double-Image Prism	215
136.	Glass Pile	216
137.	Nature of Polarisation	218
138.	Catapult	220
139.	Reflected Rod	221
140.	Refracted Rod	221
141.	Angle of Polarisation	223
142.	Quartz Plates	225
143.	Axis of Iceland Spar	227
144.	Direction of Vibrations in the Spar	228
145.	Positive and Negative Crystals	231
146.	Action of Tourmalines	232
147.	Theories concerning Common Light	235
148.	Various Kinds of Polarised Orbits	237
149.	Nicol Prism	243
150.	Large and Small Analysers	244

LIST OF ILLUSTRATIONS. xxiii

FIG.		PAGE
151.	Polariscope	246
152.	Nicol Prism Projecting Polariscope	247
153.	Foucault's Prism	251
154.	Elbow of Polariscope	253
155.	Thin Glass Analyser	254
156.	Simple Table Polariscope	255
157.	Norremberg's Doubler	256
158.	Simple Doubler	257
159.	Effect of a Crystalline Film	259
160.	Resolution of Vibrations	262
161.	Mica Designs	264
162.	Apparatus for Crystallisation	270
163.	Spring Wire Forceps	272
164.	Screw Press for Glass	275
165.	Apparatus for Heating Glass	277
166.	Effects of Sonorous Vibrations	280
167.	Fresnel's Quartz Prism	285
168.	Rotation in Quartz	285
169.	Compound Wedge Bi-quartz	289
170.	Fresnel's Rhomb	295
171.	Composition of a Circular Vibration	296
172.	Quarter-Wave Plate	300
173.	Double Mica	302
174.	Double Mica	302
175.	Mica Artificial Quartz	304
176.	Mica Artificial Quartz. Reversed Rotation	304
177.	Mica Squares	306
178.	Mica Squares	306
179.	Calcite Plate in Slider	309
180.	Vibration Planes in the Calcite	310
181.	Crystal Stage	311
182.	Tourmaline Pincette	312
183.	Convergent Lenses	317
184.	**Slide** for Mitscherlich's Experiment	323
185.	Bi-axial Wave-Shells	324
186.	Conical Refraction	325
187.	Effect of Quarter-Wave Plate	335
188.	Uni-axial Crystal Circularly Polarised	336
189.	Experimental Tube	347
190.	Multi-coloured Images	349

LIST OF PLATES.

PLATE		PAGE
I.—POLARISCOPE OBJECTS	*Frontispiece.*	
II.—THE SPECTRUM AND ITS TEACHINGS	To face	64
III.—INTERFERENCE	,,	160
IV.—INTERFERENCES OF POLARISED LIGHT	,,	264
V.—RINGS AND BRUSHES IN CRYSTALS	,,	312
VI.—CROSSED AND SUPERPOSED CRYSTALS	,,	328
VII.—ROTATORY AND CIRCULAR POLARISATION	,,	332
VIII.—SPIRAL FIGURES	,,	336

LIGHT.

CHAPTER I.

THE LANTERN AND ACCESSORY APPARATUS.

The Lantern—The Optical Objective—Gas Burners—Mineral Oil Burners—The Lime-light—Advantages of "Regulators"—Centering the Light—Mounting the Lantern—Focusing Lens—Prism—Plane Reflector—Other Apparatus—Screens—Diagrams—Vertical Lantern.

1. **The** Lantern.—Any fairly good lantern **will serve to** perform the following experiments, provided the " front " is so made as to slide on and off a flange-nozzle. This is usual with the better brass fronts made with lengthening tubes, but not with fronts half brass and half tin ; and in the latter case the alteration should be made, so **that either** the ordinary objective, or the optical objective to be presently described, or any other apparatus, will slide on and off at pleasure. The ordinary objective will be occasionally **wanted to** exhibit diagrams, while the other will be placed on the nozzle for experiments : the lantern will also be available **for** all ordinary purposes. A bi-unial is exceedingly convenient, **as the** top lantern may be used for diagrams, while the lower nozzle carries the optical objective. The usual forms of condensers are, either a pair of plano-convex lenses with convex sides turned towards each other ; or a meniscus with the concave side towards the light and a

double-convex lens in front. Either, or any form if good, will answer fairly well; and a little deficiency round the edges of the screen disc matters little for most experiments, if the *centre* be well illuminated.

2. **Optical Objective.**—The ordinary objective used for exhibiting slides would suffice more or less perfectly for many experiments; but a special optical objective is almost necessary for many, and preferable for nearly all. It is, moreover, absolutely necessary for the polariscope to be hereafter described; and, as the same lenses and fittings answer for both, and are of a very inexpensive character, it is more satisfactory to provide for efficiency at the outset. The ordinary slide-stage is also unduly large for the insertion of the necessary apertures, slits, or other apparatus, besides being inconveniently situated for manipulation; and the large field is a great waste of light for many experiments which need all we can use. Supposing, for instance, a rather small aperture has to be diffracted, we can condense no light upon it in the ordinary slide-stage; whereas by bringing it a few inches out in front, we can insert an additional convex lens, mounted in a wooden frame *as a slide*, in the ordinary stage, and so condense a very large portion of the full beam upon the aperture.

The arrangement recommended for the optical objective is shown in Fig. 1. A B is a nozzle (of japanned tin or brass, according to the style of the lantern), which slides tightly on the lantern nozzle, and is kept from rotating by a slot and pin. We will suppose it $3\frac{1}{2}$ inches diameter, for $3\frac{1}{2}$-inch condensers. B C is a 3-inch brass tube 3 inches long, which should screw into a collar at B, as it will be required to unscrew from this into the polariscope elbow, to be hereafter described. Near the bottom or nozzle end a square aperture, K, is cut through both sides to form a slide stage, which should "take" slides an inch thick, and

$2\frac{1}{4}$ inches wide. The slides are kept down to the nozzle end by a circular L-shaped collar, L, operated by studs working in longitudinal slits as usual, and forced against the slides by a spiral wire spring M, abutting against the collar at C, which screws into the other end of the tube, and has screwed into it the jacket D of the focusing tube, with its rack and pinion E. The focusing or lens tube F will be about $2\frac{1}{2}$ inches in diameter, and has screwed into it at the back end the cell of the plano-convex lens G of 5 inches focus, with the plane side towards the slide-stage. At the other end screws on a collar in which is fixed the nozzle N, projecting outwards from the front flange or collar a clear half-inch, and which callipers $1\frac{1}{4}$ inches exactly. Into the

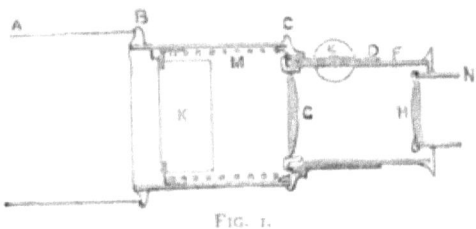

FIG. 1.

back end of this screws the lens H of 8 inches focus, which may either be a plano-convex, or of a slightly meniscus form, the whole being arranged so that the lenses are $2\frac{3}{4}$ inches apart. The focusing jacket should be so adjusted that when the tube is run right out the lens G is about 3 inches from the bottom of the slide-stage (which will focus a very small disc), and have a backward travel of about $1\frac{1}{2}$ inches; and this should be finally done by making the jacket D rather longer than necessary, so as to adjust by experiment before fitting the pinion.

It will be seen that the lenses are of a very inexpensive character; but such an arrangement is as good as can be adopted, and loses little light. It is the proper arrangement for the polariscope, for which it answers admirably; and it gives a very *flat* field with but little colour. A rack and pinion is by no means essential: indeed, if the sliding tubes are accurately circular and fit well, a plain sliding tube is preferable, in order that the lenses may be removed at pleasure and slits used with parallel light only, or other appliances (such as an adjustable slit or revolving diaphragm) slid into the tube. I have, however, found a really well-fitted plain "draw-tube" by no means so common among ordinary opticians, on this scale, as a fair rack movement. If the nozzle A B is well fitted, the objective will rarely need support; but if it does, a loose semi-circular crutch should be fitted for it, out of a thin slab of wood or otherwise. A $1\frac{1}{4}$ inch nozzle is advised because it is the recognised gauge of nearly all the polariscopes made by, or made for, the London opticians, and to which accessories and objects are adapted. It is a very great convenience to be able to obtain any such articles "to fit," and it is therefore strongly recommended that the size be, if possible, actually callipered from such an instrument. Such a polariscope can now be purchased for a very few guineas, complete, as described in a future chapter; and in that case nothing more will be necessary than to provide the additional tin or brass nozzle A B, into which the objective of the polariscope, when unscrewed from its elbow, will fit, and make the objective for straight work. The equivalent focus of the two lenses is about $3\frac{1}{4}$ inches, and this will give suitable discs for any screen distance ranging from 5 to 20 feet. If the general working distance is longer, say from 20 to 25 feet, or more, a somewhat longer focus may be preferable, making the back lens say 7 inches focal length, and giving more space between the two lenses, and between the back lens and the stage.

3. **Light.**—Sufficient effects for a class-room or moderate-sized drawing-room may be obtained in nearly all the

following experiments[1] from a *good Argand gas-burner*. A class or a few spectators do **not need to** see the phenomena on a very large scale, and by employing a small disc at a screen distance of 5 or 6 feet, very good results may be got in this way. The lantern becomes hot with such a light; but the convenience of being able to get it into work at a moment's notice, and without any apparatus, is very great. Nearly all the experiments in this work **have** been first "roughed out" with a gas-burner; and though more brilliancy may often be felt desirable, what there is **will still** be found more interesting and attractive than **any other** mode of demonstration, besides being visible **to** a whole class at once. A good "Silber" **or** one **of Sugg's best** "London" patterns may be **employed, and either** gives a very white light of from **22 to** 28 candles. **Mr. Sugg** lent me for trial a gas-burner superior to either, such as are supplied by him to the Trinity House Board for lighthouse purposes; but it **is** more costly, being 25*s*. **to** 30*s*., nickel-plated, whilst the plain Silber or London costs about 6*s*. 6*d*. In the Trinity House burner a platinum plug or button spreads the circular flame, which is again curled inwards by the throat of a contracting glass chimney; the result being a very white light of about 30 candles compressed into a space of about an inch square. Such a short flame of course focuses better than a tall one, the rays being less divergent; and with this burner excellent work may be done. It might probably be made in a cheaper form, the general construction being **a** very old one. Any gas-burner **is best** fixed in a lantern by having a slide-tray with an upright pin at the back end, as for the lime-light. Over this pin should slide a socket brazed on **a** tube, with tap and nozzle on the back end as usual, while a plain elbow at the other end carries the

---

[1] The exceptions are chiefly such as require a *parallel* beam, and are for the most part noted specially.

burner. The lime-light or plain gas can then be used as convenient, gas sufficing to "work out" privately, at a minimum cost and trouble, almost any desired experiment, even when a more brilliant light is required for public repetition of it.

The brilliant mineral oil lamps give better effects than gas, ranging in power from 50 to as high as 90 candles —that is, *standard* or "gas" candles, there being a looseness about some opticians' estimates (due to taking any candles as tests) which is not desirable. The *triple* forms of wick, of which there are several, are much to be preferred for optical work: the double wicks are apt to give a comparatively darkish streak up the centre of the screen. This is little observable in exhibiting painted slides; but in optical experiments is just where we can least afford any deficiency.[1]

4. **The Lime-light.**—The lime-light is, however, strongly to be preferred if possible, the effect being so infinitely superior, not only in brilliancy, but in *whiteness*, or completeness in the spectral colours. It is not an expensive light either, after the first purchase of the apparatus. Potassic chlorate is now cheaper than formerly, and can be purchased in most large towns at from 7*d.* to 9*d.* per pound; black oxide of manganese at about 2*d.* per pound. At these prices the lime-light for two hours will only cost about 2*s.* 6*d.* besides wear and tear; which is very little for the effect produced. The ordinary details of lime-light lantern

---

[1] Since the above was in type, Mr. Hughes, of 151, Hoxton Street, has brought out a mineral-oil lamp with *four* flat wicks, stated to give a light of 200 candles. I have not been able actually to measure its power before this work goes to press; but I have satisfied myself that it exhibits three-inch painted slides on a full twelve-feet disc, well illuminated to the edges; and—what is more to the purpose of these pages —that, with the elbow-form of lantern polariscope hereafter described, it will project either slides or crystals on a disc five feet in diameter, in a very satisfactory manner. I never expected to see this accomplished by an oil-light; and the fact that it can be will be a great aid to many country teachers and students. The heat is of course great.

management can be learnt in hand-books obtainable from any optician; but it may be well to add a few hints which are not found in any such I have seen, as well as to lay stress on the one or two points which give absolute safety in manipulation, or are necessary to secure good effects; for two operators will differ by as much as forty per cent. in the effect they obtain from the same apparatus.

The lime-light has three main forms, most properly described as (1) the spirit-lamp jet, (2) the gas "blow-through" jet, and (3) the gas "mixed" jet. The first will give from 100 to 120 candles, and is sufficient for small lecture-rooms or halls: the second will give from 140 to 200 candles: the last 350 to 450 candles, according to the pressure. All alike require a bag of *oxygen* gas, supplied to the jet under pressure; and this bag should measure 36 × 24 × 24 inches, and *not less*, for optical purposes: if there is no assistant to turn off the oxygen when not wanted, it had better be a little more for any lecture work. The size named is no more than sufficient, with a little comfortable margin, for two hours' experiment; and some is often wanted for preliminary adjustments. Only the best bags are worth having, and such will last for years if the gas is properly washed, and all taps or metal fittings cleaned and oiled every now and then.

Oxygen ought to be passed through *two* wash-bottles, if it is to be thoroughly rid of manganese-dust and chlorine. The first bottle should be fully two-thirds full, the second only about one-third full, both having a little carbonate of potassium or common washing soda dissolved in the water. The most convenient apparatus for washing will be found in common two-quart glass bottles, with loose vulcanized "Woolff" tops (like Fig. 2) to slip over the necks. In glass bottles the "speed" of the gas can be seen as it is made, and such tops can be removed and the bottles emptied in an instant. The bag to be filled should be *higher* than the last wash-bottle; and with these arrangements, only well purified and dry gas can enter. The retort is most conveniently placed over a ring gas-burner, which is amply sufficient: a good heat should be applied at first, till bubbles *begin* to come over, after which it should be slackened, and only moderate heat will be required. This management of the heat is the only way to get the gas over with fair regularity.

The precautions in making oxygen are chiefly, to be sure the pipe from the retort is disconnected from the first wash-bottle *before* the retort is taken off the fire or gas (otherwise water may be sucked back and cause an explosion), and to be sure no organic substances are in the mixture. The safety tube will do the rest, and the india-rubber tops on the wash-bottles are additional safety-valves against a sudden rush of gas. The chlorate should be spread abroad on a sheet of cartridge paper to see if any small bits of straw, stick, or other matter have got in; and a small portion of the black oxide should be mixed with chlorate and heated in a test-tube.

FIG. 4.

Practically the only danger here is, lest a portion of soot or charcoal might be mixed with the sample; and if this is found pure, it is as well to lay in sufficient manganese from the same lot to last the winter.

Making hydrogen is troublesome, and is not to be recommended wherever house gas can be obtained. It is a common idea that it gives a better light, but this is quite an error; very careful experiments purposely made having failed to show any perceptible difference. If it has to be made, particular care must be taken to have all common air driven off before any is collected, and also all air expelled from the bag.

There is not very much variation in spirit jets; **but** blow-through jets differ widely, and are certainly capable of improvement. The prevalent fault **is** too large an opening for the house gas, which heats the lantern very unnecessarily. The best of the ordinary forms is probably that shown in Fig. 3, the oxygen nozzle o being somewhat below the gas nozzle H, giving a better mixture, with a smaller and more pointed flame. Such a jet should be selected with the outer aperture as small as possible, and an inclination of about 45 degrees.

With the "mixed" jet no accident can occur if the pressure from each gas is fairly equal, and *no pressure is then altered* while the jet is alight. There is no practical danger in *adding* weight on a double pressure-board permanently. With the blow-through form, 28 lbs. on the bag at first will **be** enough, gradually increased to 56 lbs.; and many lose light as well as waste gas by commencing **with too** much: but the mixed jet gives the best results with high pressure. Unfortunately, however, the pressure does not remain uniform, but diminishes as the bags empty, the result of which is that a fresh adjustment of the taps is necessary every now and then. A valued correspondent found **by** experiment that a pressure equal to 9 inches with full bags and taps closed, gradually diminished to little more than **4** inches as the gases exhausted: also that it **was very** difficult to get the same pressure with both gases, whatever arrangement was adopted. He therefore introduced between each bag and the lantern a simple gas regulator,[1] and then found that, provided **there** was only *sufficient* pressure behind,

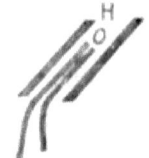

Fig. 3.

[1] Any efficient regulator ought to answer for this purpose; but the small and handy form used by my correspondent, and since by myself, with most satisfactory results, is manufactured by Messrs. Parkinson and Co., Cottage Lane, City Road. They cost 3s. to 6s. each.

it could be kept uniformly steady, and the same adjustments could be retained throughout. In using these regulators, the pressures are first tested as usual, by U-tubes like Fig. 4. The tube is open at one end, while the gas to be tested is led to the other by a vulcanized stopper and small L-pipe. A wooden slab between the branches has a line ruled across about the middle as a zero point, above and below which it is divided on opposite sides into inches and tenths. Water is poured in to the zero point; and when the gas is turned on, it of course depresses one column and raises the other, the *difference* indicating the pressure. Any variation in weight will tell at once on this index: but it will be found that when the regulator is introduced, and adjusted to a given pressure, that pressure will never vary so long as a pressure not inferior is kept up behind it. Both gases can thus be adjusted equally to any pressure which the weights at command will enable the bags to keep up to the end. With such regulators, and trial with the U-tube, if there is a tolerable pressure of gas from the main, the "mixed" jet can thus be used with *only one gas-bag*, which is a very great advantage, giving even at the pressure of the main (say 2 inches, or 1 inch on each side of the U-tube) a better light than most blow-through jets. Another advantage the "mixed" jet used in this way has over the ordinary "blow-through" jet, is that the oxygen gas lasts nearly double the time, or goes twice as far. A single jet may thus be employed at all times, either with gas from the main to give with half the oxygen a light equal to that of the "blow-through" form, or, with an extra bag and more pressure, the full light of the mixed gases.

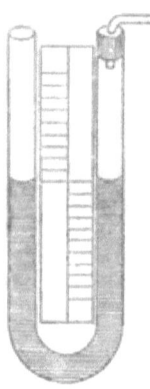

FIG. 4.

The lime has to be brought very close to the **mixed jet** —say $\frac{1}{8}$ of an inch from the orifice: hence the inclination and a well-finished conical point are of importance. Many jets are so clumsy and so low in angle, that they cast a shadow on the **condensers,** "standing **in their** own light," as **it were. For optical** work it is important this should **not** be **the case. To** avoid it, **some** makers place the jet upright; the cylinder, or flat **side** of a disc of lime, being itself inclined at **45°**. This is, however, dangerous, the heated gases being reflected **off** the lime direct to the condensers. In **any case** the lantern should be "warmed up" cautiously to avoid cracking the lenses.

The limes well-known to opticians as "Excelsiors" are far the best. "Soft" limes are often recommended **for** blow-through jets, but those named **give at** least equal, if not greater, brilliancy, with a smaller radiant spot; **and** with a "blow-through" require hardly any turning. With the mixed jet the lime should be turned a little, the same way, *with every experiment*. Finally, the best result will depend on the supply of each gas being carefully and deliberately adjusted; and the greatest advantage of the regulators referred to, is, that when this has once been effected it can be retained with certainty throughout the experiments, leaving the demonstrator or his assistant **free** (beyond turning the taps on and off as required) to give attention to other things.

5. **Centering the Light.**—When the lime-light is employed, it should be adjusted in the axis of the optical system of the lantern with much more care than is usually taken over slides. The best way is to make a cap of black card to fit over or into the condenser cell, with an aperture about $\frac{1}{8}$ of an inch in diameter accurately in the centre. Remove all the lenses, place this cap over the cell, and a smaller similar one, or the smallest hole of the revolving

diaphragm, if there is one, as far in front of the nozzle as can be managed. Bring up the lime as near the cap as is practicable without burning it, and then adjust with care till the spot on the screen shows the greatest possible size and brilliancy. Very much, in some experiments, depends upon this, and on keeping the lime properly turned. We may very often see fine experiments, with the most costly apparatus money could purchase, spoilt by a dark patch upon the screen, and thus worse rendered than a careful operator would have done with a plain "blow-through" jet. This latter, well managed, is in fact amply sufficient for nearly all purposes. With lamps, very accurate centering is of less importance, owing to the larger luminous surface.

Fig. 5.

Whatever the light employed, we term the original source, in the lantern, the "radiant."

6. **Mounting and Using the Lantern.**—This is very conveniently done upon a small revolving wooden tripod stand like Fig. 5. The top circle turns upon a centre-screw passing through the under one, by which it can be tightened from beneath in any angular position, the total height being from 6 inches to 10 inches, and the lantern fixed to the top in any convenient manner. This being placed on a good-sized table, the lantern is conveniently raised, and easily turned in any direction, while the large table is available for accessory apparatus.

If prism-work forms a very important feature in the experiments, it is better to extend the revolving top on one side by a *board* extending out in front (of course with a supporting leg) about 3 feet, furnished with a cross-board which can be slid backwards or forwards to any distance from the lantern. The advantage of this plan is, that when a slit or anything else has been properly focused and the rest adjusted, the whole arrangement can be deflected off at any desired angle for refraction or reflection, without having to readjust the apparatus.

For optical work all lenses ought to be in brilliant order. They should only be cleaned with a perfectly clean (well beaten and shaken when dry) wash-leather, kept in a clean box for that purpose. If the leather alone fails to rub them bright, a little alcohol applied with cotton-wool will remove the dulness. A damp "fog" will often appear when the lantern is first lit, and must be allowed gradually to disappear before anything can be done.

7. **Accessory Apparatus.** —What is needed for special experiments will be described in the proper places; but some accessories will be in almost constant demand. First of all

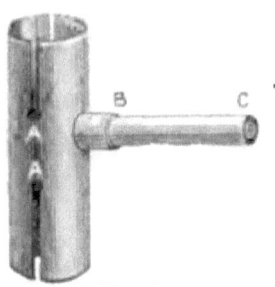

Fig. 6.

will be required two pillar or rod-stands, made by screwing or casting a half-inch brass tube neatly plugged at the top into a heavy foot of lead or iron, which should have baize glued on the bottom to prevent slipping. Such can be bought at gas-fitting shops for three or four shillings each. On these should slide closely the sockets shown in Fig. 6, into which are screwed or brazed the small tubes B C, which receive the various pieces of apparatus. As it is sometimes necessary to get well *over* some other article, there should be a few of these, with the length B C various; and if the weight and leverage are too great for the stand, another socket bearing a leaden weight as counterpoise will keep all steady. In the socket tube itself, small holes, A A, are bored, and saw-cuts made down to them, when a slight pinch together of the two semi-circles so divided, at each end, will make an admirably tight *sliding* fixture, independent of screws or any such nuisance.

Into these sockets we can fit anything. The loose focusing lenses, which will be constantly needed, can be fixed by three short pins, as at A, into a circular hole, turned with a ledge in a disc of wood, B (Fig. 7), into the edge of which is driven a short length of tube, C, fitting into the socket tube B C (Fig. 6). Or if there is no lathe at command, the hole may be cut with a fret-saw, and the lens fixed with short pins each side. It can of course be mounted in brass if preferred; but in that case it is well, when using, to put round it a border of black card, to exclude from the screen all but the rays focused. A lens of 3 inches diameter and 7 inches focus is a good sort, and a meniscus, with the concave side to the screen, is to be preferred where the expense of an achromatic lens is an object; but two lenses of 6 inches and 10 inches are very handy, according as we may want to enlarge and collect all the light we can from a small object, or to focus a large one. Quarter-inch tube is a good size for these accessories, but a gauge should be chosen which can always be matched, and the socket B C (Fig. 6) made to fit. In fact, a foot or two of tube to mount any apparatus should always be at hand.

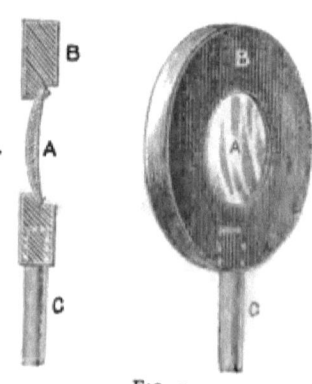

FIG. 7.

ACCESSORY APPARATUS.

A reflector, say **6 inches by 4 inches, of good** thin looking-glass, may be **mounted as in** Fig. 8. D B **is a** length of **the** tubing, D C fitting into the socket. From A to B half the tube is filed **away, with a notch** at C, **into which one long edge of** the glass fits. The B **end is** still **further** filed away; **and when the** glass **A B is put in** position, with **a strip of card or blotting** behind and **round the bearing edges, to impart**

Fig. 8

a little "give," the end B **is brought over, and** B **and** C **gently pinched** down. The half-tube will keep **it nice and** stiff. A **reflector may be also** mounted **in other ways, and for some experiments a second one** is handy, **but seldom really required.**

A **glass** prism, 2 **inches** long, and **about** 1¼ inches in face, **of good** dispersive power, can **be bought** for about

Fig. 9.

4s. 6d., **and is** mounted **as in Fig. 9,** the tube being fixed into a triangular cast socket, **or one made** of sheet brass turned **up** ¼ inch. **Into** this **the end** of the prism is cemented. Of course, such a prism **is** not "optically" **faced,** only polished **like** lustres; but it is good enough for most **screen work.** In the socket, Fig. **6,** it can be **set** and **turned to** any angle horizontally; **but another** socket should be provided, as in Fig. 10, bearing a piece of brass, A B, at the end, in which is bored, perpendicularly, a hole of the regular socket gauge. In this the prism can be turned to any angle vertically.

For spectrum work a prism bottle filled with bisulphide of carbon is, however, far to be preferred, giving nearly double the dispersion of any ordinary glass prism. Such a prism can, however, only be used vertically, and a glass one is handy for some other positions sometimes necessary.

These fittings are recommended because they can be adjusted in work with such ease, precision, and rapidity,

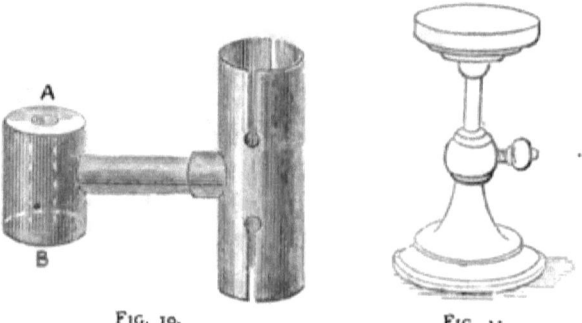

FIG. 10.   FIG. 11.

and are so readily interchangeable in all sorts of combinations. In these respects they are far superior to the usual method of mounting focusing-lens, prisms, &c., each on the top of a special stand, besides being much cheaper, and easier made.

An adjustable table-stand somewhat resembling Fig. 11 will be needed for many things. The handiest size for average work, with such a tripod as Fig. 5, is one which will rise from 12 to 18 inches, with a top about 6 inches across. Some kind of adjustable clip will also be needed to hold wires, cards, plates, &c. The best construction for all purposes is what is known as the "Bunsen universal holder," shown in Fig. 12. The clip itself can either be inserted vertically into the pillar-socket, or the intermediate joint will allow of any angle or other variety of position. The price of each of these, if purchased, will be about 6s.

A cardboard screen about a foot square should be provided, though it is not actually necessary. It is easily and cheaply made by screwing or glueing an inch wooden rod or ruler into any wooden foot, sawing down the middle from the top about three inches, and sticking the edge of a thick sheet of cardboard in the slit. The centre should be the same height as the lantern objective from the operating table; and if one side of the card is white and the other blacked, it will serve either to stop light when required, or to receive an image.

Various blackened cards or thin zinc plates, each 4 inches by 2¼ inches, will also be required, in which to cut suitable apertures for insertion in the optical stage. The zinc also should be blackened if used.

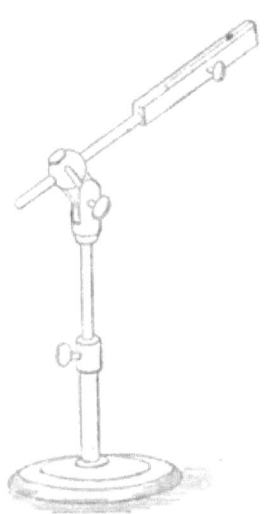

FIG. 13.—Bunsen Holder.

An excellent optical black varnish may be made by mixing "vegetable" black at discretion with a mixture (about half and half) of ordinary French polish and methylated spirit. On small work it should be applied with a large camel's hair or other soft brush; on large woodwork with a paint brush. For this varnish is suitable for almost every purpose, from blacking cards to large gas-bag pressure-boards. It dries quickly a full dead black, and does not come off, or warp the cards.

8. **Screens.**—Where there is a good white plaster wall, well smoothed, nothing will surpass it. Next to that, for small screens, the best is a single sheet of white paper. At many large printing-offices they now use sheets 60 × 45 inches; and probably any one who really wanted a few sheets for such purposes would be courteously obliged. They can be fixed to a wall with drawing-pins. Large screens are made of linen faced with paper. Transparent screens both spoil the more delicate effects, and cause a tremendous loss of light. It is, however, impossible to avoid them in some places; but the operator must in such cases be content with a small disc, in order to avail himself of the only tolerable material, which is fine tracing *cloth*—the fine varnished muslin kind. This can be procured anywhere 42 inches wide, and as transparencies are usually wanted in small space, this is amply large enough. Such cloth shows infinitely better than the best wetted sheet, and there is no need whatever of stretching it on a hoop. It is just as well to mount it on a roller at top and bottom, and such a screen may be fixed in a minute by tying the top roller to anything, say two branches of a chandelier, and attaching a small weight to the bottom rod or roller. Transparent screens are not, however, as already observed, nearly so well adapted as an opaque white surface for optical work, especially with the prism; and it is moreover often desirable that an audience or class should see, not only the effect upon the screen, but the experimental means by which it is produced.

For some experiments we shall also want a horizontal screen over the apparatus, and for this a white ceiling will answer perfectly. We shall have to employ it almost immediately.

9. **Diagrams.**—These can be prepared in many ways, but the following are the best. White diagrams on a black

ground may be done with **stout** needle-points (either **compass** or ruling points), through photographic black varnish spread thinly and evenly, when firm, but not quite so dry as to be chippy. Or a glass may be warmed and smeared with solid paraffin, which is then cleaned **partially off and re-melted** so as to be as thin and even as possible; **if the plate** is then held over the smoke of burning camphor this will adhere, and though not so hardy as the other, will **bear** careful usage, and cuts easier. Compass **curves are best** struck from small bits of thin horn as centres. These engraved diagrams are very striking; but the **darkness of the screen** sometimes hinders necessary details being pointed **out, and** makes black lines on a white **ground** preferable for many. There are two good methods for **these also; the** first being to draw with a hard pencil upon the *very finest* ground glass —a process any one can execute **at once;** the other is **to** *scratch* with needle-points on a sheet of gelatine, rubbing **a** little blacklead into the scratches, and mounting **the** film between two clean glasses. Or a drawing may **be** made on gelatine with Indian ink rubbed **up** till black enough. **All** diagrams should be faced with glass when completed, and bound with strong dark gummed paper. This edging paper must be gummed very freely and *dried*, the strips being moistened when used. Wet gum, freshly applied, does **not** adhere to glass with certainty.

10. Vertical Attachment.—Though only **needed** for one or two experiments mentioned in this work, and those not essential ones, **a** vertical attachment capable of projecting horizontal surfaces is so generally useful in acoustic projections and for other purposes, that a description will be of use. As invented and perfected by Professor Morton of America, it is rather an elaborate piece of apparatus; but all required can be done very simply and cheaply as shown in Fig. 13. A cubical wooden box is open on one of the perpendicular sides, and has a circular aperture cut in the top nearly extending to the edges of the square. The size of the box and of this aperture depend on the size of field to be projected; for magnetic curves or wave-ripples, a circle of less than five or six inches diameter is of little service. Opposite the open part, **on strips of wood at** each side, a piece of **good plate looking-glass is supported** at an angle of 45°, so as to throw the horizontal beam from the lantern up perpendicularly through the circular aperture. Owing to the larger size of this (**the horizontal field**) a *diverging* beam **will generally be** required, **from any ordinary lantern, to cover it; and hence there** *should* **be a large plano-convex lens (costing about 15s. for one 6 inches in diameter) fixed to the under** side of the aperture, so as to re-converge the rays on the focusing lens. Fair effects **can, however, be procured without this with a good lime-light, in spite of what is then wasted. In any case, all but the condensers are removed from the lantern, and** these are so adjusted, and the

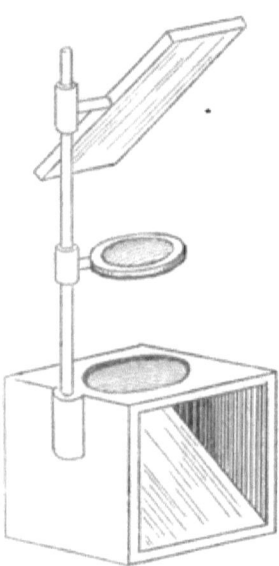

FIG. 13.—Vertical Attachment.

box so placed in front, that the field is just covered and no more. To one side of the box is fixed a stout perpendicular brass rod or tube, the same gauge as the pillar-stands already described, and furnished with two sockets like Fig. 6. **The lower socket** bears a focusing **lens, either plain or achromatic**—it **may** be the one already described (Fig. 7)— and the upper socket the **plane reflector, which** reflects the image to the screen. These two sockets must, of course, be fitted to the apparatus, as it is necessary that the lengths B C (Fig. 6) should in this case be such as to bring the lens and reflector central with the perpendicular axis; but the lens and mirror already described will answer for the rest perfectly well. It might be supposed that the first surface of the mirror would cause a double image on the screen; but it is so faint in comparison with the other, that any such effect is rarely perceivable at all, and only then to close and special observation.

Besides its other more strictly experimental uses, the vertical attachment is very effective for projecting on the screen *extempore* diagrams, which are easily traced on a sheet of prepared glass laid horizontally. It often makes a demonstration much clearer to work out a somewhat complicated diagram step by step in this way. The ground glass and pencil is best adapted for this method of work, as the operator can work more freely and see more precisely what he is doing.

CHAPTER II.

RAYS AND IMAGES.—REFLECTION.

Rays of **Light**—**Rays** form Images—Inversion of Images—Shadows—Law of Intensity at various distances—Law of Reflection—Virtual and Multiple Images—The Doubled Angle of Reflection—Application of this in the Reflecting Mirror—Reflection from Concave and Convex Mirrors—Images **produced by Concave** Mirrors—Scattered Reflection—**Light** Invisible.

11. **Rays of Light.**—All objects that are visible to us, are so in virtue of "rays of light" (whatever these may prove to be) which proceed from every point in the luminous surface. That these rays are perfectly straight or rectilinear while traversing the same medium, observation convinces us. We can trace the straight path of a sunbeam in a dusty room; we know, or find by experiment, that light from a lamp can only be seen through three perforated cards arranged with an interval between them, if all three apertures are in a perfectly straight line; and we are conscious that if any opaque body be held in the straight line between our eye and a lamp the light is effectually stopped. But the rest of the mechanism—the real *image-forming* power of these rays—is masked to the ordinary observer by the number of them. To see it clearly, we must isolate the rays which proceed from the object in certain directions.

12. **Images.**—We may do this by darkening a room and

only allowing light to enter through a very small hole in the shutter—the original *camera obscura*. Bringing a sheet of white cardboard near the aperture (Fig. 14), we see depicted

Fig 14.—Image formed by a small hole.

upon it the landscape outside, and we thus learn experimentally that rays of light are really sent out from all points of visible objects, and that these rays form images.

This is equally the case whether the objects are self-luminous, like the sun or a lamp (with which class of bodies we more commonly associate the idea of this ray-sending property), or whether they only reflect to us light derived originally from other luminous sources. In all cases the *image* is formed on the retina or on a screen, by rays which proceed from the object to the site of the image.

This landscape image is rather faint, because the objects depicted are not very brilliant, and there are too few rays passing through the aperture to form a bright one. We can get a brighter image by collecting wider cones of rays from each point of the object, methods of doing which will speedily come before us; or the more brilliant light of the lantern will yield a better one. Remove all the lenses, including the condensers,[1] and cover the flange-nozzle with a piece of tinfoil or a cap of black card. In this prick a hole with a rather thick needle,[2] and an image of the radiant at once appears on the screen. Since each point of this image is defined by straight lines proceeding from the corresponding point of the radiant through the prick to the screen, it is obvious that if we prick another hole we must get another image. We go on pricking holes, an image appearing with each, till by degrees the opaque tinfoil or card is removed. As we do so the images of the radiant crowd on one another more and more; presently they touch;

---

[1] A simple and striking experiment for the private student is to take off the top, and knock out the bottom of a coffee canister, and blacken it inside. Then punch a hole in the middle of its length, about one-sixteenth of an inch in diameter. Hold this over a naked candle in an otherwise dark room; and a good image of the candle-flame will be formed at six or eight inches' distance upon a white card or finely-ground glass.

[2] A needle-prick will give good images from a lime-cylinder. With lower illuminants, the prick will have to be made larger, and a sheet of card, or the portable screen, should be brought within a few feet of the lantern.

then they overlap; at last, when all the tinfoil is gone, the screen is covered with a white glare of light.

13. **Necessity for Isolating Phenomena.**—We now see how the image-producing power of the rays proceeding from any luminous object is ordinarily masked by the mere multiplicity of such rays. We say the screen is "lighted," but we have found that this illumination really consists in its **being** covered over by an infinite number of images of the radiant in the lantern. We thus learn at the outset a lesson of the greatest importance, viz., that in many cases, to ascertain the true nature of the phenomena of Light, we must *isolate* a comparatively small number of rays. Unless **we** do so,

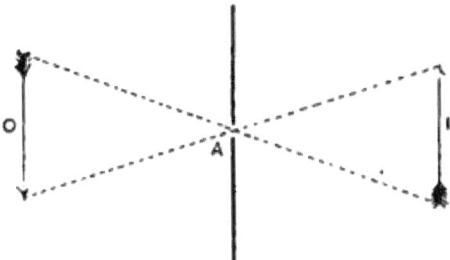

FIG. 15.—Inversion of image.

such a number of images (or other phenomena under examination) may get mixed up, or partially superposed, that their real character may be lost in a general confused body of light due to them all.

14. **Inversion and Relative Size of Images.**—It is plain, also, why such images as we have been forming must be *inverted*. Let o, Fig. 15, represent the original object; then it will be seen that the rays from its top and bottom, or right and left, if they proceed in straight lines, must cross at the aperture A, those from the top proceeding to form its image I at the bottom. It is further clear that the relative sizes of

image and object must be exactly proportional to their respective distances from the aperture **A**. Both these truths **are of** general and important application in practical optics.

15. Rays, Beams, and Pencils of Light.—We have thus far dealt generally with " rays," but we soon perceive that a single ray must be in reality a mere ideal abstraction— *one* precise mathematical **direction,** or line without thickness, out of many divergent rays from every luminous point. To get an image, as in Fig. 15, from single rays, the aperture A must be *infinitely* small, but under this condition visible phenomena would vanish altogether ! In practice we **can** only deal with whole bundles of rays, the larger of

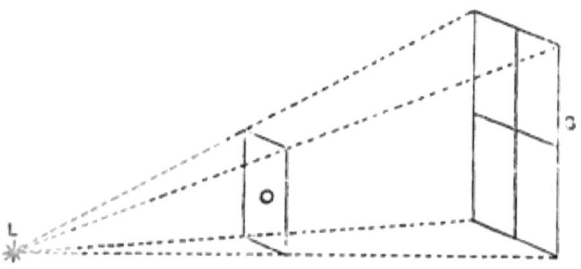

FIG. 16.—Shadows and Law of Squares.

which are usually called "beams," and smaller ones (whether parallel or conical) **"pencils."** If, however, we clearly understand this, **it is** convenient, in colloquial **language, to** speak of a small parallel pencil as **a ray, and the word is** often used, though not with **strict correctness, in** this sense.

16. Shadows.—That rays of light proceed in straight lines **as long as they traverse** the same medium, explains **why opaque bodies cast** *shadows.* Fig. 16 **makes** this clear. It **is plain that if the square of** card (s) is held between the naked light **in the** lantern and the screen, it **must cast a shadow over the section of** the whole **cone** of rays which it intercepts. It is plain, also, that while a small bright point of light must cast a sharp shadow, a large source of light must cast one fringed by more or less partial shadow called a penumbra; and if the source of light be larger than the intercepting object, at a certain distance behind the object there must occur a position where **the** whole of the light is never intercepted, and no total shadow exists.

17. **Law of Inverse Squares.**—Fig. 16 also illustrates the law of the *intensity* of light. If the object, o, is exactly halfway between L and S, its shadow will cover on the screen a space exactly equal to four similar squares. At twice the distance the light which reaches o has to cover four times

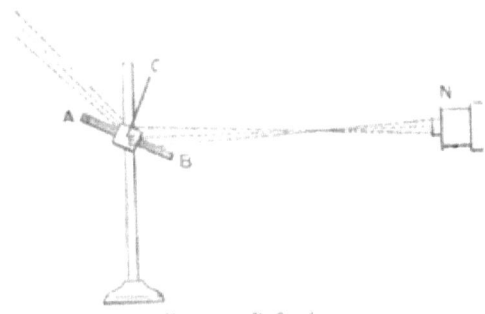

FIG. 17. - Reflection.

the area; at thrice the distance, **nine times.** Hence the intensity of light is "inversely as the square of the distance" from the radiant point.

18. **Reflection.—Law of Equal Angles.**—We must now replace the condensers in the lantern and put on the optical objective, placing in the stage a blackened card or zinc plate 4 × 2¼ inches, with a horizontal slit in its centre an inch long and about ⅛ inch wide.[1] Three or four feet in front

---

[1] With the lime-light, if a nozzle with adjustable slit is at command, the parallel beam through this, without any objective, is better. But either arrangement will answer.

of the nozzle, N, place one of the pillar-stands, with the plane reflector, A B, in one of the horizontal sockets, as in Fig. 17. Fasten on the edge of the reflector just opposite to the socket, by a groove or wide saw-cut, a piece of cork, E, in which stick, at right angles to the reflector, a wire, C, as an index. The slit turned horizontally will project a thin slice of light (as it were) against the reflector, and it is at once seen that if this impinges at right angles it returns on its own path; but that, if the ray be incident at any angle with the perpendicular index, it is reflected at the same angle, as shown in the figure, on the other side of that perpendicular. A little smoke from a bit of lighted touch-paper[1] will make the course of the beam perfectly evident. Owing to the fact that any angle of the incident ray is thus reduplicated on the other side of the perpendicular, while the perpendicular incidence alone is reflected in the same path, this perpendicular is called the normal, and angles in optics are reckoned from the normal, and not from the actual reflecting surface. Thus the polarising angle of glass is called 56°, and not 34°, which is the actual angle with the surface.

This fact, that the angle of incidence is equal to the angle of reflection, was about the first optical law known to the ancients, and has several important consequences. First of all, it leads us to a grand general principle of *reversibility*. That is, if the radiant were removed to the "other end" of the ray, the path traversed would be precisely the same, reversed. This is very generally true in practical optics.

19. **Virtual Images.**—Secondly, we perceive that a polished or perfect reflector must produce "virtual images." We "see" things in the direction from which the image-forming rays last come to the eye. Hence, taking the rays from a bright source of light, such, for instance, as a candle-flame (see Fig. 18), the divergent rays, when reflected, *appear*

---

[1] Soft paper dipped in a solution of common nitre.

as though they proceeded from a point behind the mirror, at which therefore there appears a "virtual image."

20. **Multiple Images.**—As the eye can see the object direct, and also its virtual image, the object appears to be duplicated; and it is easily seen that if more than one mirror be employed, images may easily be further multiplied. Two parallel mirrors, by successive reflections, can be made to give a number of images, a familiar example of which may be found by looking at the image of a candle-flame in a piece of thick looking-glass held obliquely; a

FIG. 18.—Reflected image.

whole row of images will be seen, as in Fig. 19, owing to the rays being reflected and re-reflected between and from *both* surfaces of the glass. At a great angle of incidence, the reflection from the first surface of the glass is as bright, or even brighter than that from the silver surface; showing again that the comparative intensity or completeness of reflection from various substances, when polished, differs with the angle of incidence. If a piece of plain glass, instead of looking-glass, be used, we see also that reflection takes place, not only when rays encounter the polished surface of a denser medium, as glass; but also at the second surface, where the rays meet the rarer medium of the air. We shall hereafter find that this last kind of reflection is often the more brilliant of the two. In all cases, reflection is the more copious the greater the angle of incidence, except in this last kind of reflection.

Repeated reflection sometimes produces most beautiful effects. When two mirrors are inclined together at an angle which is any aliquot part of a circle, or 360°, and

FIG. 19.—Multiple images.

the rays from any object pass in a path between them, nearly parallel to the junction, there must be as many images (including the object) as the angle is contained in 360°. A glance at Fig. 20 will show how this occurs. Two such mirrors, or even plain rectangular oblong strips of plate glass, fixed in a tube, with a cap at one end made of two parallel transverse glasses, between which are loosely contained some coloured beads, or other transparent objects, and with a small hole in the other (closed) end of the tube, form the *kaleidoscope* of Sir David Brewster. In Darker's Lantern Kaleidoscope, the mirrors are of platinized glass, and are mounted with a convex lens at each end; thus mounted the apparatus takes the place of the ordinary lantern objective, and will produce beautiful patterns upon the screen if a rotating slide containing the fragments of coloured glass is placed in the ordinary slide-stage. In using this instrument the light must be *raised* half an inch

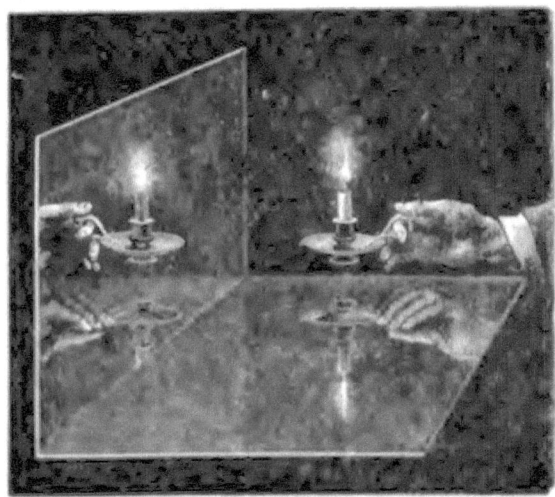

FIG. 20.—Symmetrical multiple images.

to an inch above the usual central position, and the angle of the mirrors turned downwards. This is to ensure that all the light which reaches the screen is reflected down upon, and upwards from, the mirrors; and the effect will resemble that in Fig. 21.

21. **The Doubled Angle of** Reflection.—Another very important consequence follows from this law of equal angles. It is obvious that in changing the position of the

mirror, the reflecting surface itself moves through the same angular distance as the index or normal. Hence it follows that any angular movement of the mirror is *doubled* by the angular movement of the reflected ray; and this fact makes the "reflecting mirror" an invaluable method of recording minute motions. We are familiar with the lever-index, moving from a centre; but in practice we are fettered in this means of multiplying small motions, by the weight and

FIG. 21.—Kaleidoscope.

other mechanical imperfections of an index-pointer of great length. In the reflected ray of light we not only double the angle to start with, but we have a pointer we can make of any length, which is absolutely straight, and weighs nothing at all. Hence the "reflecting mirror" has constant applications, of which the following will serve as experimental examples.

THE REFLECTING MIRROR.

22. The Reflecting Mirror.—By keen sight, in the right light, the motion of the pulse may just be discerned, though it is almost imperceptible. But cut a piece of looking-glass an inch square, and paste on its face a bit of black paper with a circular hole $\frac{1}{2}$ inch in diameter. To the back attach in a triangle three pellets of wax, or anything that will stick to the skin, and stick the little mirror on the wrist, with one of the pellets just on the pulse. In the slide-stage place a zinc-plate or card with a $\frac{1}{4}$ inch circular hole, and focus the reflected image of the aperture. Hold the wrist in the beam, so that the incidence is about 45°, as in Fig. 22. At once the motion of the pulse is made visible to all by a motion of the reflected

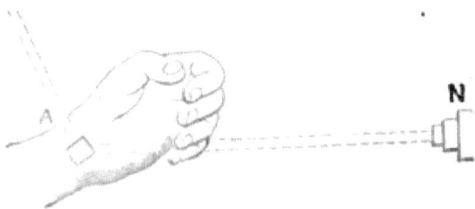

FIG. 22.

spot of light on the ceiling of several inches. This is a very pretty and striking experiment, though simple. In preparing for it, it is well to find the exact spot on the wrist at leisure, and to mark it by a dot of ink.

In the same way we may demonstrate the rapid and minute motions produced by heat. Upon the table we adjust a Trevelyan rocker, A, with its block of lead, L, and fulcrum-knob as usual. Having heated it, we fix on its face, by any cement that will bear the heat, a small mirror B, like that just used, and by our glass reflector, C, direct the beam of light from a small aperture down upon it at any angle, so as to be reflected to the ceiling or the screen,

D

where the spot **should be** focused sharply. The whole arrangement **is** shown in **Fig. 23**. What would **be** a **stationary spot of** light if the **rocker** were cold, is by **the** small rocking motion at once prolonged (by the persistence of impressions **on the** retina) into a bright line of light, **and** by gradually raising with the hand the block **D** bearing the fulcrum **end, this is** converted into a beautiful wavy line E F, which **makes** every separate motion **visible.** The rocker should be judiciously **chosen by** experiment as to its period **of** vibration. A kitchen poker **may** be made to

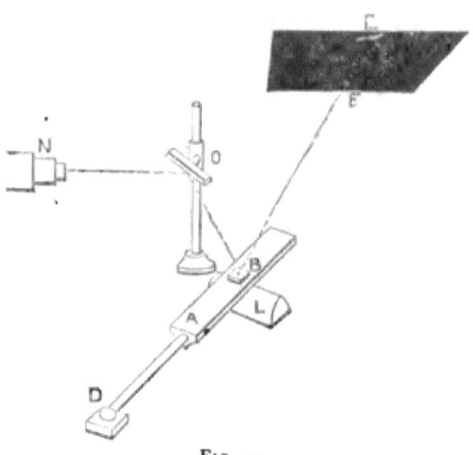

Fig. 23.

do by bevelling **one side of the square end to a very** obtuse angle, and attaching **the mirror to the opposite** face.

Our next illustrations **may be from** those vibrations which ·**pr**o**duce sound. The** elegant experiment of M. **Lissajous may** easily be illustrated **with the** lantern and **such a tuning-fork as may be bought for 5**s**.** On the outside **of the end of** one prong must be strongly cemented **a** small bit **of** looking-glass, **and** on the other **a bit of** metal **or** glass **to** balance it. **We** then mount the fork in a

THE REFLECTING MIRROR.

heavy block of lead, place a slide with a small hole in the slide-stage, and arrange the whole with the plane reflector in the vertical socket, as in Fig. 24, so that the light from the lantern may be reflected back from the mirror on the fork A to that on the pillar-stand B, and thence to the screen, where it produces a spot, which must be focused.[1] The card screen should also be placed between the fork and screen, with its blackened side towards the lantern, so that none of the incident beam may pass and interfere with the

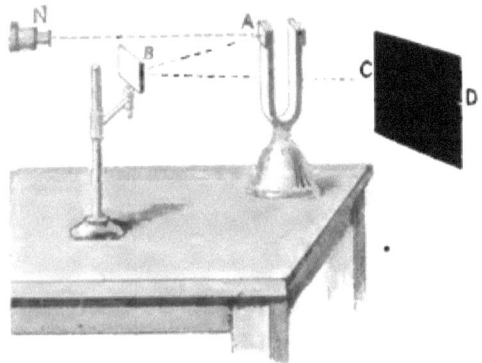

FIG. 24.

effect. On now exciting the fork by a violin bow, the spot is expanded by the angular motion of its mirror into a

[1] These are the usual arrangements for a gas-burner; but where the lime-light is used, it will often be preferable so to place this accurately in the condenser focus as to make the whole beam parallel, to remove the objective altogether, and place the proper apertures on the front of the nozzle, as is usually done with an electric lantern. A rough tin cap is easily made in which the zinc plates can slide. Such parallel beams are not focused. In all cases the mirror used should be as large as practicable, as the brightness of the focused spot will depend on the size of the pencil of light reflected by the mirror. These remarks apply generally to the class of experiments depending upon reflection from a small mirror.

bright vertical line ; and **by** slightly turning the reflector ʙ
in its vertical **socket, this is** developed, as before, into a

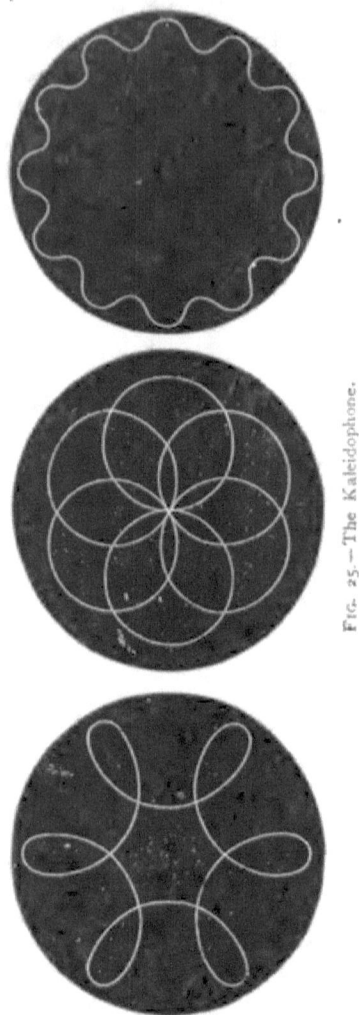

Fig. 25.—The Kaleidophone.

beautiful undulatory form **c** D, showing each vibration of
the fork. **We** may, **as is well** known, substitute a second

fork fixed horizontally for the mirror B, and thus get beautiful compounded curves; but this belongs rather to another subject, and a pair of really accurate forks are expensive.[1] Equally beautiful *optical* effects may be got by squaring off carefully the end of a springy steel rod, cementing a small mirror on that,[2] and fixing the other end firmly by a screw into a mass of metal sliding on one of the pillars. This is presented "end on" to the nozzle of the lantern, in the position of the fork-mirror A in Fig. 24, but the reflecting mirror B is not moved. On now drawing aside the rod, and releasing it, or striking it, beautiful curves will be produced, of which Fig. 25 may serve as specimens, depending on the thickness and length of the rod, where it is struck, &c. This is the simplest screen adaptation of Sir Charles Wheatstone's *Kaleidophone*. Mr. Ladd has made a modification of it called the *Tonophant*, in which the rod is made of two steel slips of different lengths welded together like a girder; and by bringing more or less out of the screw socket, almost any of Lissajous' combinations can be produced.

Get a large and thin claret or champagne glass—the larger the better—say 3 inches diameter, and fill to the brim with alum-water. Adjust this on the table as in Fig. 26, the reflector A throwing the full light from the nozzle N at a

[1] Forks that perform very well optically, may be bent up of steel about 1 × ¼ inch. The arms should be about a foot long, and each arm furnished with a metal socket sliding easily on it, and fixed by a screw at any point. By these movable sockets we have much control over the periods of vibration. Such forks need not be polished.

[2] These very small mirrors are not easy to fix and to shape, even of the thinnest plate-glass obtainable. Minute bits of glass silvered on the first surface, after the manner of specula, do better. The most convenient material of all I have myself found to be small polished discs of very thin steel, such as are sometimes used to ornament feminine articles. Unfortunately they are not easy to procure. Those I have used "came off" a fan.

slight angle down on the glass, and the lens B focusing the surface on the ceiling.[1] On now ringing the glass by the edge **of a** knife, circular **waves of** light and shade will be seen: and on dipping the finger in the alum-water and—holding the stem firmly down—rubbing the edge till **the** well-known **sound** is produced, small ripples will cover the surface in exquisite **patterns,** and follow the finger round, all being reproduced above on **the ceiling.** With a large thin glass the pattern may **often be varied by** pressure, but glasses

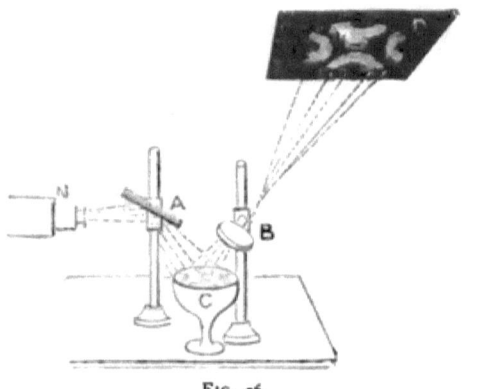

FIG. 26.

differ. A violin-bow, besides being in the way, produces somewhat too strong vibrations.

[1] If an Argand burner is used, **we** want all the light, and **the** reflector and glass must be so brought up near the nozzle, that all is collected just into the circumference of the water. All scattered light must also be carefully stopped ; covering the side-openings of the ordinary slide-stage with a cloth, seeing no light escapes anywhere, and standing the **glass itself on a piece of** black cloth, that no light may be reflected from **the table surface to the** ceiling. In working with **a** low illumination much depends on these precautions, and with **them this beautiful ex**periment will show very fairly on **a** ceiling 9 or 10 feet high ; **a reflector** in the **lantern is also of service.** With the lime-light **we** need not be so **particular, and the parallel beam** will be best.

THE REFLECTING MIRROR.

The vibrations **due** to sound may be shown by the reflecting mirror in yet another way, due originally to Professor Dolbear. Procure a tube of any material about $1\frac{1}{2}$ inches diameter, and say a foot long—paste-board will do, or metal. Over one end of it stretch any thin membrane —part of **a** child's india-rubber balloon will answer excellently, or even a piece of paper gummed on. In the centre of this fix by **a** little gum or other cement a small bit of thin looking-glass,[1] **not** exceeding at most $\frac{1}{4}$ inch

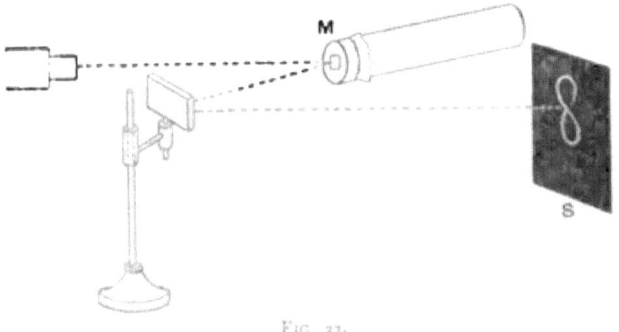

Fig. 27.

square. **The whole is to be** arranged **in any** kind of crutch **or rest as in** Fig. **27, reflecting** a **spot** of light to the screen from the small **mirror** and **plane** reflector. Then sing into the open end of **the** tube. **Every note** will produce **a** line or figure of some kind, each line varying in azimuth, and the figures often taking symmetrical forms, as the membrane vibrates under the sonorous vibrations in the tube. This simple experiment is very interesting and beautiful. Still more beautiful methods of optically representing these sound-waves will come before us when we consider

---

[1] Thicker glass is apt to separate too much the image from the first surface. See also note on page 37.

the colours of thin films. Meantime these experiments may suffice for simple plane reflection, and the power of the reflecting mirror; which last finds its fullest development in the "rotating" cubical mirror of Sir Charles Wheatstone, by which the velocity of light itself was measured by Foucault with such marvellous accuracy.

23. **Reflection from Curved Surfaces. Concave Mirrors.**—Further consequences follow from the law, or fact, that the angles of incidence and reflection are equal. Let us suppose our reflecting surface, instead of being flat, is curved, as in a portion of a sphere, of which a section is

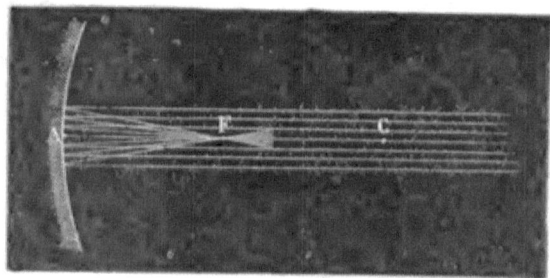

FIG. 28.—Concave Mirror.

shown in Fig. 28, the centre of the curve or sphere being at C. A drawing upon a large scale will speedily show that a series of parallel lines, representing rays, leaving the mirror A at equal angles to those of incidence, must meet or converge *nearly* at the point F, midway between C and A. If we take a divergent pencil as in Fig. 29, we find still the same thing: here C is still the centre of the mirror, from which all lines drawn to it are perpendiculars to its surface. One such normal is shown by the dotted line, and simple inspection makes it evident that the lowest ray from S, reflected from the mirror at an equal angle on the other side of that dotted line, proceeds **to** *s,* and that **every** other ray from s is reflected to nearly the same point. The qualification "nearly" **is** necessary, because **a** carefully constructed mathematical diagram will show that only a *parabolic* form will exactly converge parallel **rays** to one point, and that a spherical mirror only exactly converges rays emanating from the centre of the sphere. But for small surfaces the aberration is small even with **spherical mirrors ; and** these **are the** correct figure for **aiding the** light in **lanterns** or electric **cameras, the** rays proceeding from the centre of curvature, and being reflected

FIG. 29.—Image formed by Concave Mirror.

back through the same point, so as to reach the condensers at exactly the same angles **as the direct light.** Parabolic mirrors **are** often fitted **by** opticians **to lanterns;** but a moment's reflection will show that such **an** arrangement is a mistake, **as** the condensers cannot deal properly at once with **the** divergent light from the radiant and the parallel **light** from the mirror. On the other hand, when a strong parallel beam is required, as in lighthouses, the parabola is the correct form. **The** point where parallel rays converge is called the principal **focus.**

24. Images from Concave Mirrors.—As the rays from one point are collected by a concave mirror to another point, as in Fig. 29, we must necessarily have an image (§ 12). And because a wide cone of rays is thus collected and converged, this will be a *brighter* image than those previously obtained. We have supposed s to be a point in Fig. 29, but if we take an object A B (Fig. 30) and trace only two diverging pencils for the sake of clearness from its top and bottom points, we shall see that both rays from A converge to *a*, and both from B at the point *b*; parallel

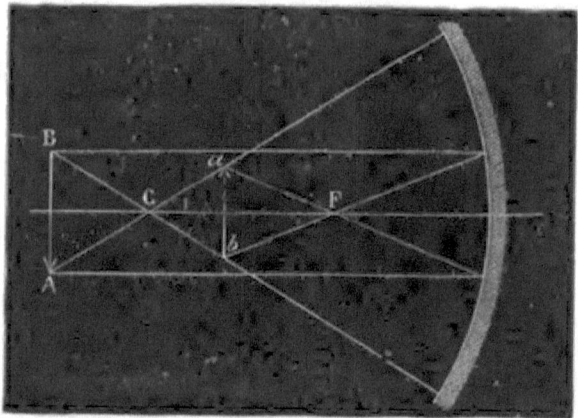

FIG. 30.—Inverted Image.

rays crossing at F, the principal focus of the mirror, and rays perpendicular to the surface at C, the centre of curvature. If *a b* is the object, then the enlarged image will appear at A B, owing to the principle of reversibility (§ 18). Inspection will show that the image must be inverted.

All this can be easily proved by experiment, a concave silvered glass mirror 6 or 8 inches diameter being now procurable for a few shillings, and answering the purpose sufficiently,[1] the secondary image from the first surface of the glass being too faint to matter much in mere demonstration. We simply remove the objective from the lantern, deflect it about 90° from the screen, and place in the ordinary lantern slide-stage any diagram, or some simple lantern-slide. The concave mirror (of not less than 6 inches focus) can then easily be so held by hand as to throw a very fair image upon the screen of the slide in the lantern.

This, then, is one method of collecting sufficient rays from an object to form bright images; which is practically utilised in the *reflecting telescope*, largely used for astronomical purposes. The rays to be collected being practically parallel, the figure of the mirror is parabolic; and the collected rays may be brought to the eye in either of several methods, which need not be described in a work dealing only with the physical phenomena of light.[2] If the pupil of the eye be, say $\frac{1}{8}$th of an inch in diameter, and we "see" a star by the "light" which enters that small area, it will readily be understood how many thousand times the same quantity of light will be collected by a good speculum several feet in diameter, and how much magnification an image so produced will bear without too much loss of brilliancy.

25. **Convex Mirrors. Virtual Images.**—The images just considered are *real* images: that is, the rays diverging from all points of the object being actually converged to certain points, if a screen be adjusted at these points the image or picture will appear upon it. A few moments' consideration

---

[1] For private experiment all the phenomena of curved mirrors may be easily traced out by blacking the convex side of a watch-glass for the concave mirror, and the concave side of another for the convex mirror.

[2] These and other optical instruments are fully explained and illustrated in Guillemin's *Applications of Physical Forces*, published by Macmillan and Co.

and a simple diagram [1] will show that if an object be placed in front of a convex mirror, the divergent rays from it must be reflected still more divergent, and must, therefore, since the rays are "seen" in their last apparent direction, *appear* to proceed from a point behind the mirror, obtained by prolonging behind it the reflected ray-lines. Such an image appears smaller, and erect, and is a "virtual" image, having no real existence. If an object be placed nearer a concave mirror than its principal focus, a diagram will also show that no real image can be formed, but that a "virtual," erect, enlarged image will appear behind the mirror.

26. **Scattered Reflection.**—Turn the lantern again towards the screen, throwing from it either a beam of parallel light, or the image of an aperture an inch diameter if a plain burner is employed, stopping the beam with the black card screen at the end of the table. With the plane mirror throw the beam back, and rather towards the ceiling, to any point not very white or light-coloured. Nearly the whole of the light will be reflected, but the mirror itself will be little illuminated, and the room itself will remain nearly dark. Now take the card screen, and turning round its *white* side, use that in place of the mirror. The beam of light is no longer reflected as before (in the form of a bright beam) to the ceiling; but the card is brightly illuminated, and a very considerable amount of light is diffused throughout the room. Hence the light at first sight appears to be reflected according to different laws in the two cases.

But it is not really so, and this simple experiment, with what we have already discovered about forming images by collecting and converging diverging cones of rays, explains

---

[1] The student is strongly advised to construct such diagrams *for himself*, solely by the method of drawing ray-lines at equal angles of incidence and reflection.

to us how *things* become visible, or send light to us, though that light be only borrowed or reflected. We can understand in a moment that a *perfectly* polished surface, if such were possible, can only reflect to us or converge for us, the diverging rays from other and luminous objects, without altering them otherwise. It must itself, therefore, be utterly invisible. Such a perfect polish is unattainable, but our nearest approaches to it prove the truth of this. It is not uncommon for a large mirror occupying the whole of one end of an apartment, or the side of a landing, to be unperceived;[1] and Colonel Stodare's "Sphinx," which made some sensation years ago, depended upon the space between the legs of an apparently three-legged table being glazed with brilliant looking-glass. The stage being kept in a rather subdued light, and the carpet and hangings carefully arranged of uniform colour, with no "pattern," to all appearance there were only three legs under the table and box which supported a living head; whereas the man's body was comfortably accommodated behind the two opaque mirrors, placed with their angle of junction towards the spectator.

With surfaces not perfectly polished, or—which means the same thing—not perfectly *smooth*, it is very different. Let Fig. 31 represent such a surface, with its inequalities highly magnified. The rays from the left hand, say of a sun-beam, strike upon it all parallel. But even the few which for the sake of clearness are drawn, are reflected in all directions, owing to the furrows and protuberances of all sorts, which make the angle of incidence variable for almost every one. It can be seen that every reflected line in the diagram is drawn at the proper angle; yet how different these reflected directions are! In reality the variety of directions

[1] I once walked up against a large mirror placed at the corner of a club staircase, and occupying the entire wall at the corner.

is countless, and thus from every sensible point of the surface comes, not the original parallel pencil of rays, but a *cone of divergent rays*. The body itself thus behaves like a candle, or has become luminous, though only by reflected light; and thus it is that if we collect and converge these new cones of rays, either by the eye, or by any other methods, we form an image.

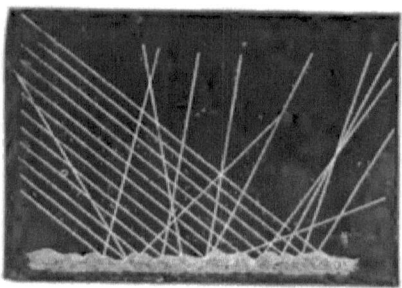

FIG. 31.— Reflection from Unpolished Surfaces.

27. **Light Invisible.**—We push our experiments on scattered reflection a step further, for it bears on a very important matter. Already we cannot help asking ourselves, what *is* this Light, which obeys rigidly such simple laws, and yet produces such various effects by them? The equal angles of reflection and incidence almost irresistibly suggest to us a ball rebounding from a wall, or a billiard-ball from a cushion, which (except for the influence of any "twist" in the latter case) obey the same law. It is natural to conceive of Light as consisting of infinitely small and highly elastic particles, propelled from the original luminous source; and such a hypothesis will account for most of what we have found, if not all. Being very simple and convenient, and so easily understood, we may therefore provisionally adopt it; but besides some other difficulties we must grapple with later on, the hypothesis has one obvious difficulty that

encounters us even now. Light ought, if it be as supposed, to be *visible*. And indeed we are apt to picture it as possessing intrinsic brilliancy of its own, and we have even appeared in many of the previous experiments to "see" the course of the **rays—the** very "rays" themselves—in our darkened room.

Nevertheless it **is not so, as** a careful consideration of our last experiments **soon leads us to perceive. This Light we** are studying is not itself a Thing, but a **Revealer of** things. It **is** itself, **and** by itself, absolutely *invisible*. It *makes* visible **to us** luminous objects or sources, rays from which actually **reach** our eyes ; but if we look "sideways" at rays from the **most** dazzling light, we cannot see them. Space is black. If **we** appear in previous experiments to have " seen " **the course of** the rays in our darkened room, this is only because of the little motes in the air ; and Professor Tyndall has shown that, destroying these by **heat,** and keeping fresh **ones** out of a glass tube thus cleared, the **space traversed by the** full beam of an electric lamp **is dark as** night. We demonstrate this less perfectly, but sufficiently, as in Fig. **32.** Place on the table a confectioner's glass jar, A, 6 inches in diameter, cover it with a glass plate, B, and drop into it a bit of smoking touch-paper, which soon fills the jar with smoke. Adjust the plane reflector, C, to throw the whole light down as parallel as possible, when the jar is **at once** filled **with a** peculiar lambent light. Take off **the** plate and **let the smoke** out ; and, as it disappears, dark spaces appear where there are no particles to reflect the light, till all is dark. The Light itself is there alike at all times ; but where there are no solid particles to reflect it actually *to the eye*, we see nothing at all. The Light that illuminated the jar is itself invisible.

Once more : clear the jar and fill it with clean water. Again it is almost invisible, except where the rays may be reflected from some point of the glass direct to the eye ; and it would be quite invisible were the clearness of the water and polish of the glass perfect, as we have seen (§ 26). But now pour in one or two spoonfuls of milk and stir it up. At once a splendid opal light fills the jar, and a pleasant radiance the room.

Many students, and even teachers, too much despise these more simple experiments; but they are not only of great beauty, they are pregnant with meaning. We have here not only a difficulty in the "emission" theory which we must not forget, but we have had another striking example of scattered

FIG. 32.—Scattered Reflection.

reflection—that kind of reflection by which most bodies are seen. In white light, more or less of such scattered light is always white. Therefore "coloured" bodies reflect white light as well as coloured; and more white light will be reflected from a black hat in the light, than from a shirt-front in the shade. Smoke is soot, and we all know soot is black; but in our last experiment but one, the light we got reflected from our particles of smoke, when diffused, was white.

CHAPTER III.

REFRACTION.—TOTAL REFLECTION.—PRISMS AND LENSES.

The Refraction or Bending of Rays—The Law of Sines—Index of Refraction—Total Reflection—The Luminous Cascade—Prisms—Lenses—Images produced by Lenses—Focus of a Lens—Virtual Images and Foci.

28. **Refraction.**—Provide a rectangular tank about two inches between the sides, one of which, to serve as the front is a piece of glass a foot square, or a little more; let one end also be of glass, and the top open.[1] Paint over the face with black varnish all but a circle, on which paint a horizontal and perpendicular line through the centre, as in Fig. 33. Provide also a strip of thin zinc or copper blackened, C D, rather wider than the tank, and about three inches longer, in which cut two slits $\frac{1}{8}$ inch wide, and nearly the whole width of the strip (or depth from front to back of the tank) in length, in such positions that when the strip rests perpendicularly against the glass end, the slit E shall be about $\frac{1}{2}$ inch above the horizontal line, and the slit F make an

---

[1] Professor Tyndall was the first to employ this striking method for lantern demonstration of refraction. His tank was circular, like a clock-face. The square form here described is more easily made, and the loose cover and slits have advantages in demonstrating total reflection.

E

angle of 40 degrees from the centre of the circle with the horizontal line.

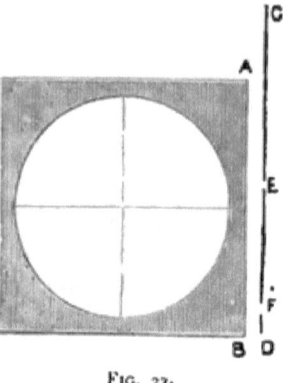

FIG. 33.

Fill the tank exactly to the horizontal line with water mixed with two or three drops only of milk; place the metal strip over the top with both slits towards the lantern, and arrange the reflector as in Fig. 34, placing in the optical

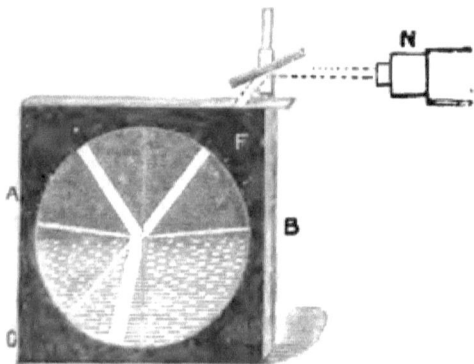

FIG. 34.—Refraction and Reflection.

stage the slit used in our first experiment in reflection, horizontally, or using it with parallel light. First of all direct the light through the slit E (Fig. 33), only a little off the perpendicular, covering over the other slit. It will be seen that the ray is bent or refracted on entering the water. Cover up this slit and uncover the slit F (Fig. 34), near the end of the tank, and blow in a very little smoke from a bit of touch-paper to show the course of the ray. (If too much is used the light will be scattered, as in our last experiment.) We now see more clearly all that takes place. First of all, the ray is much *more* bent than before; and secondly, a considerable part is *reflected* according to the law found in our former experiments.

29. **The Law of Refraction.**—We have found that in passing from the air into the water the ray of light is wrenched or bent down towards the perpendicular, or *refracted*, as it is called. The greater the original angle with the perpendicular the greater also is this bending, and we naturally inquire if there is any law which governs these variable angles of deflection. The law is simple enough when known, but not very obvious to mere observation. It eluded even Kepler's special investigation, and was only discovered about 1620-25, by Willebrod Snell. It is called the "law of sines." Taking a circle, as drawn on the face of our tank, described round the point where the ray enters the denser or more refractive [1] medium, and drawing the perpendicular or normal, A B (Fig. 35), we may take any incident ray, D C, and the refracted ray, C d. From the points at which these cut our circle we let fall D S and d s perpendicularly upon the normal A B; then D S is "the sine" of the angle of incidence, and d s that of the angle of refraction, and the two lines will bear a certain proportion; in the case of air and water here supposed, it is almost precisely as 4 : 3. Now take any other angle of incidence, E C, and its refracted ray,

---

[1] A heavier fluid may have less refractive power, or *optical* density, than a lighter one. Oil of turpentine floats on water, but has much more refractive power.

c e, and drawing the sines as before, they will bear *precisely the same proportion.* All the sines bear the same invariable ratio.

30. **Index of Refraction.**—When this ratio of the sines is put into the form of a fraction—generally a decimal fraction —it is called the " index of refraction." Unless otherwise specified, figures so given are understood with reference to *air* as unity. In the case of air and water we have seen that this ratio is *nearly* $\frac{4}{3}$; and when put into decimals, 1·335 is the "index of refraction" for water. It follows, that the greater the refractive power the higher the index must be.

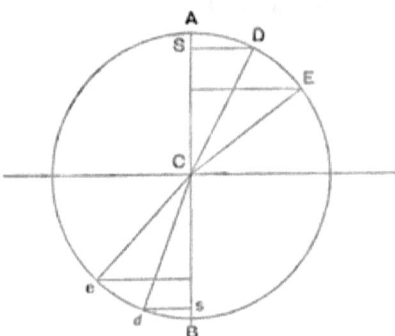

FIG. 35.—The Law of Sines.

31. **Refraction into a Rarer Medium.**—We have proved that a **ray passing** obliquely from air into water is bent towards the perpendicular, or downwards; and yet if we look at a stick standing in clear water it *appears* to be bent upwards. Fig. 36 explains this. The dotted lines represent the **real** position of the bottom part of the stick, and those **dotted** from the lowest point show the course of the rays which reach the eye from that point. On reaching the surface they are bent *from* the perpendicular, and the **bottom of the** stick is "seen" in the direction **from which the rays** actually enter the eye.

We thus see that the course of the refracted ray, like that of a reflected ray, is exactly *reversible*. If the bottom of our tank is made of glass, and it is raised up from the table and a ray sent *up* through the water, it can be shown experimentally that at sufficient angles the ray is refracted from the perpendicular on leaving the water: or the ray may enter the top of the tank as before, and after first being refracted downwards, will, on passing through the glass bottom, be again refracted *upwards*, or from the normal.

FIG. 36.

32. **Total Reflection.**—But there is a curious limit to this. Seeing that as the **angle** of incidence increases, the refracted ray is more and more bent downwards, or towards the perpendicular, we cover the top of the tank with **a bit of** plain board, **and place the** metal **strip upright against the end** of the tank, in the position **of Fig. 37.** Gently **canting** our lantern a little, **we** pass the **beam** direct through the slit E (Fig. 37), **so as to enter the water almost** horizontally. We find the ray **bent down a** great deal, about at **an angle of** 45°, **as** shown by **the thin** white line C D. **Now we have** ascertained that **if we** throw the ray first **by our reflector**

up through the water, in this case the path is exactly *reversed*. It occurs to us at once, that if we sent our beam

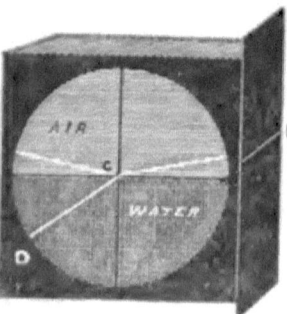

FIG. 37.

at a slightly greater angle (never forget that all angles measure from the normal) through the water, there is no path in the air it can assume : it would appear that it could not get out of the water at all. We try it, as in Fig. 38, sending our beam up through the lower slit by a bit of looking-glass, so as to strike the centre at an angle of 50° from the normal.[1] It does *not* get out—not a sign of it. It is totally reflected ; and because this reflection is total, it has a brilliancy not possessed by that from even the best silvered mirrors. It will

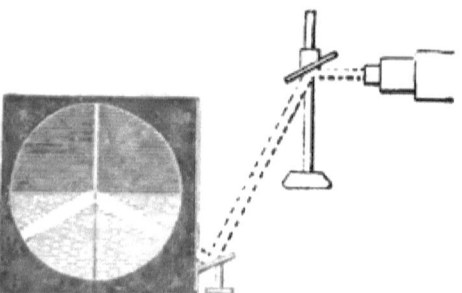

FIG. 38.—Total Reflection.

be readily seen that the angle of total reflection must *decrease*[2] as the index of refraction *increases ;* but this will be

---

[1] The limiting angle is about 48°, but we go on the safe side.
[2] Never forget that these angles measure from the normal. The phenomena of total reflection may be observed by the private student by looking at the under-surface of the water in a tumbler held rather

shown by a beautiful experiment when **we come to** study the subject of colour.

33. **A Luminous Cascade.**—There is another very beautiful method in which total reflection may be illustrated by the lantern, called the "luminous cascade," or "fountain of fire," **which may be** arranged so as to be very effective by simple **means. Get a** two-necked glass receiver (Fig. 39) about $4\frac{1}{2}$ **inches** diameter, with as large necks as possible, and in each neck fix by **corks** glass tubes of similar size, as large as possible, not less than $\frac{3}{8}$ inch clear bore, and $\frac{1}{2}$ inch is **better.**[1] Black varnish all outside, except a circle, **c,** three inches diameter, opposite **the** neck meant **to be** horizontal, **and** adjust this as at **A** (Fig. 40) close against and projecting into the lantern nozzle, (the flange nozzle, with the objective removed,) **on any** stand, filling with water first, and corking the tube in the horizontal neck till all **is** arranged. Several feet higher, fix some sort **of** supply tank (a bucket will do) with a bit of tube fixed by a **cork** in the bottom, and **connect** with the top neck by a

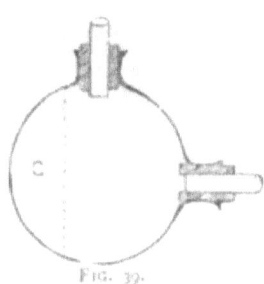

Fig. 39.

above the level **of** the head and of a candle, **or by immersing the** lower **end of a** test-tube slantwise in water. Whatever **is placed in** the dry **tube will** be invisible, and the tube brighter **than** silver; but **on** pouring in water the brightness disappears and the **contents** of the tube become visible.

[1] There is sometimes difficulty in procuring **a** receiver with both necks large enough, on a small globe. **In** that case, insert as large a tube as it will take in the largest neck, for the emission opening, and strain the flexible supply tube over the other neck alone, which will give plenty of aperture for the supply. A fair-sized stream is a *sine qua non;* otherwise there is not enough light, and it breaks up too soon into drops.

flexible tube, B, the whole arrangement being shown in Fig. 40. Finally adjust the light at such distance from the condensers that the greatest possible amount is concentrated into the space occupied by the emission-nozzle. Having adjusted all this, and filled the tank, remove the cork from the tube, and let the water stream out in a gentle curve into a bucket on the floor. The effect is beautiful even on

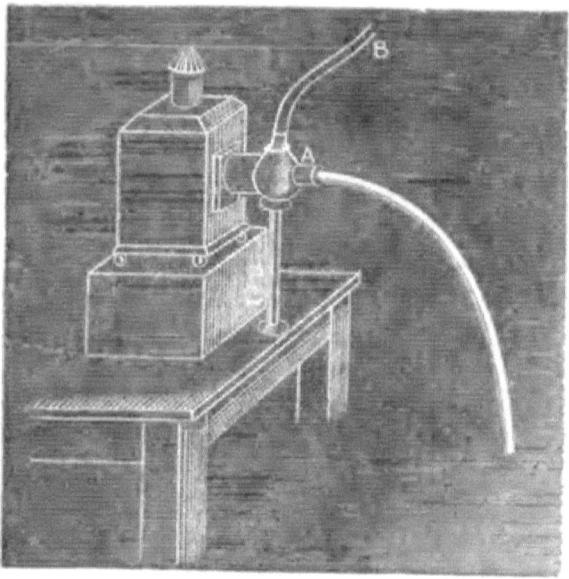

FIG. 40 —Luminous Cascade.

this small scale. The jet is like a stream of living fire ; and if we have some coloured glasses and slip them in turn into the ordinary slide-stage of the lantern, we get blood-red, blue, or what colour we desire. All this is owing to " total reflection." If the water did not issue, and we replaced the cork by a ground-glass stopper with flat polished ends, we know the light from the lantern would be thrown horizontally into the room. But it meets the stream of water on every side at much more than the angle of total reflection; and so it cannot get out, but is reflected from side to side all down the stream, making it brilliantly luminous by the small motes in the water. Put the hand in the jet, and it is bathed in light—that light which cannot get out of the stream except where we thus break it up.

34. **Deflected Rays.**—We must, however, follow refraction a little further. We have seen that the path of a refracted ray is reversible, and that, on leaving the water, it is bent *from* the perpendicular. We easily see (Fig. 41) that

FIG. 41. Deflection.

if a ray, s, passes from and into air through a denser medium with parallel surfaces, as a thick piece of plate-glass, it must be deflected somewhat, but finally resume a direction *parallel* to the original, as if proceeding from s'. We may demonstrate this with the lantern, by placing in the ordinary stage with objective removed, and focusing on the screen with the loose lens, a blackened glass slide with one or more perpendicular lines, or other figures, scratched in white through the varnish. Now hold across the slide a strip of plate glass ¼ inch thick, so as only to cover a portion of the figure. When this is held flat to the slide there is of course no perceptible refraction ; but when the plate glass **is held** obliquely, so that one end is farther from the slide than the other, **the** lines as far as covered are perceptibly deflected **or** broken. (Fig. **42**.)

35. Prisms.—But we further perceive that if the **two** surfaces of the glass are inclined **to** each other, the ray must be permanently **deflected into a** new direction. Any refracting body with **faces so inclined is** called **a** prism. **Fig. 43** shows a section of such a prism, **and it is** clear that a ray, S I, infringing **on the first surface at the angle** S I N with the normal N, will **be refracted in the path** I E **towards** the perpendicular. But E N′ **is the normal to the second** surface,

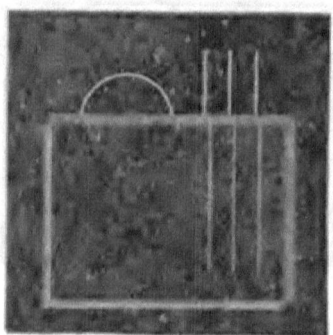

FIG. 42.—Deflected Image.

and on **emerging the ray must be refracted** *from* that, in the direction E R, widely **different from that of** the incident ray. It is equally clear that **the deviation** must depend not only on the density, **but on the angle of** the refracting surfaces of **the prism.** It also, however, depends **on** the *position* of the prism (§ 36).

All this is easily demonstrated ; but to avoid much colour **phenomena,** which must be studied separately, it is best to take a water-prism of rather a small angle. Such a prism is readily **constructed** by shaping a **smooth** wedge of beech about 3 inches square, one inch thick at one side, and tapering to an edge at the other. Bore centrally through, from face to face, a circular hole 2 inches diameter; paint the inside of the hole with sealing-wax varnish; and then, heating two clean pieces of plate glass 3 inches square, cement them with hot sealing-wax or shellac on the flat sides. By a small hole bored from one end of the wedge this prism may be filled with water, or the hole may be entirely opened out to one end so as to form a small open trough. Placing a small aperture in the optical stage, and focusing on the screen,

FIG. 43.—Prism.

stand the water-prism on one end on the table-stand (Fig. 11) and adjust in the path of the rays. It will at once be seen how the image is deflected. If a multiplying glass, which can be bought for a shilling or two, or the button from a glass "lustre," is held in front of the nozzle, a number of refracted images will appear on the screen.

36. **Position of Minimum Deviation.**—It will soon be found that the deflection varies as the prism is turned round upon its axis into different positions. It will also be found that the deflection or refraction is least when the prism is placed as in Fig. 43, so that the incident and refracted rays make *equal angles* with their respective surfaces. This

position is known as that of "minimum deviation," and is carefully arranged for in all accurate prism work, such as spectrum analysis.

37. Lenses.—Let us now consider a number of prisms of gradually increasing angles, arranged round an axis. Fig. 44

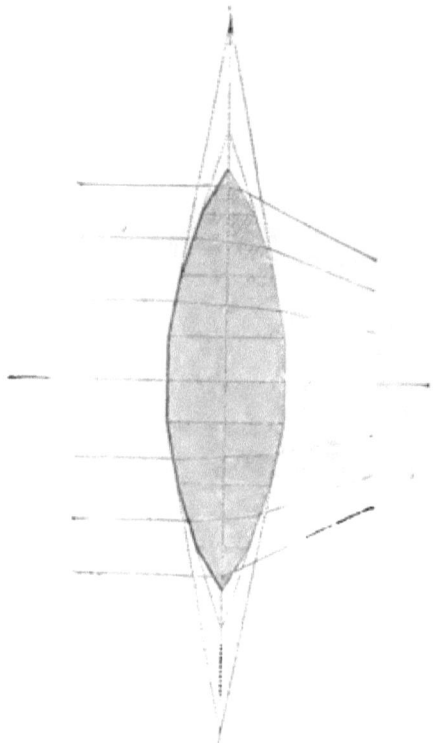

FIG. 44.—Nature of a Lens.

may represent such a combination of prisms in section. A glance at the diagram shows that the outer parallel rays, from the left hand, meeting prisms of greater obliquities, are more deflected than those nearer the centre; and that if the obliquities are properly adjusted, all might be made to converge in one point. If the obliquities are infinite in number, it is obvious we get *curved* surfaces, and such form a *lens*. These may either be convex or concave on either or both surfaces, or flat on one, and convex or concave on the other. It is plain that whatever the figure may be, if they are thicker in the middle than the circumference they will be *converging* lenses ; if thickest at the circumference, *diverging* lenses.[1]

38. **Images formed by Lenses.**—Since parallel rays, falling on the double-convex lens A (Fig. 45) converge to the

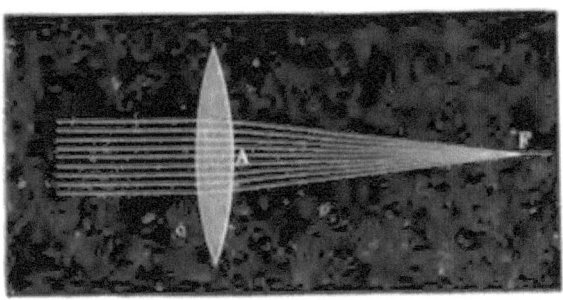

FIG. 45.—Lens and Focus.

point F, which (as in the case of a converging mirror) is the principal focus ; and since the path of the rays is reversible, and rays from the point F after traversing the lens become parallel, a moment's consideration will make clear that if the rays diverge from any point *beyond* the principal focus, they must converge to some other point and form an image.

[1] A lens formed by two equal convex surfaces is called a double convex lens, and if the two convex surfaces are of different curves, a "crossed" lens ; lenses with two concave surfaces are double-concave lenses; those with one side flat, plano-convex or plano-concave ; a lens with one side convex and one concave, a meniscus.

Fig. 46 shows this. The *parallel* rays from A and B converge at the principal focus F after traversing the lens O; but if rays *diverge* from the points A and B of an object (only pairs of rays are shown for the sake of clearness) they converge to the points *a b* and form an image. It is also clear how the respective distances of image and object govern their respective sizes, so that if *a b* is the object, A B will be its image. Also that the image thus formed must be inverted.

Thus we have a second method of forming brilliant images, the lens taking up a large cone of rays from each

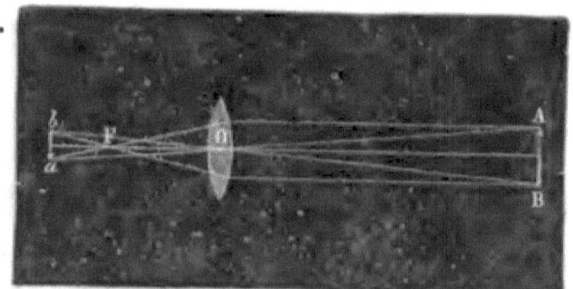

FIG. 46.—Image formed by Lens.

point of the object. As in the case of mirrors, however, the spherical surfaces which are most easily ground do not truly converge the rays to a point, except for a small central portion of the lens; the figure necessary to do this being parabolic. Such parabolic lenses have been ground, though with great difficulty. There are however other errors also to be corrected; and it is easier and more convenient to correct all these errors by methods presently described, or to stop off some of the most erroneous marginal rays.

39. **Virtual Images and Foci.**—A mere inspection of Fig. 47 will show that if the object or luminous point

be nearer the convex lens than its principal focus F, the rays cannot form an image, but simply become less divergent. If the emergent ray-lines are produced back to s', that will be the "virtual" focus. In this case we have

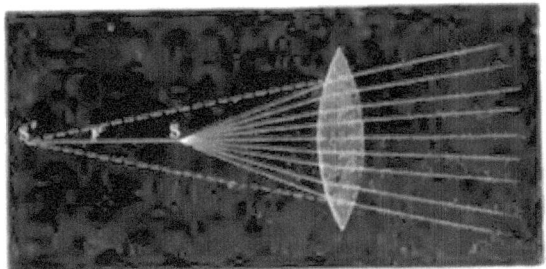

FIG. 47.—Lens and Virtual Image.

also a virtual, magnified and erect image, as in the ordinary way of using a magnifying glass.[1]

40. **Concave Lenses.**—A double concave or other diverging lens either converts parallel rays into divergent, as in

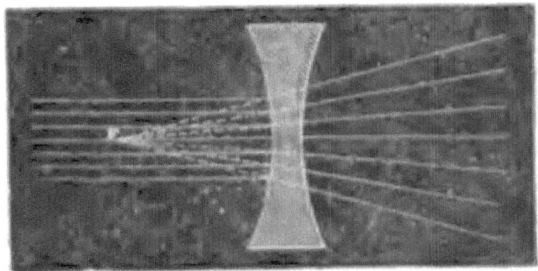

FIG. 48.—Concave Lens.

Fig. 48, or convergent rays into *less* convergent ones, which may be parallel, or still remain convergent. Such lenses

[1] The student is again strongly urged to work out for himself diagrams of this and other cases; not here described in detail, as unsuitable to the experimental character of this work.

can only have a "virtual" principal focus, F, obtained by producing back the divergent ray-lines into which they refract parallel rays.

A whole host of optical instruments, which cannot here be described,[1] are based upon these properties of lenses; especially microscopes, **telescopes**, and such **lanterns as we employ in our** experiments **or for** exhibiting views. In most **microscopes and** telescopes, a magnified image is further magnified by an eye-piece. In the lantern, the "condenser" lenses are employed to make the diverging rays from the radiant parallel or nearly so; while the objective, or our loose focusing lens, forms a magnified inverted image upon the screen of some object powerfully illuminated. Except as regards details necessary for correcting aberrations, these phenomena of refraction explain them all; and every curve of every lens has to be calculated for the special work that lens has to do, according to its index of refraction and the law of sines.

[1] See Guillemin's *Applications of Physical Forces.* Macmillan & Co.

Pl. 2.   THE SPECTRUM AND ITS TEACHINGS

A   Spectrum of Incandescent lime.
B   Wide slit, with its spectrum, showing white centre
C   Ditto   Ditto   analysed.
D   Spectrum of the sun. E Absorption Spectrum of Sodium
F   Emission Sodium Spectrum. G Absorption Spectrum of Chlorophyll.

'Wright's Light'

CHAPTER IV.

DISPERSION AND THE SPECTRUM.—DIFFERENT COLOURS HAVE DIFFERENT REFRANGIBILITY.

The Spectrum—Different Colours differently Refracted—and each Colour has its own Angle of Total Reflection—Position of the Prism and its Effect—Correction of Aberrations by Variations in Position—White Light a Compound of Various Colours—Suppression of Colour produces Colour—Artificial Composition of White Light—A Narrow Slit necessary for a Pure Spectrum—The Rainbow—Refraction and Dispersion not Proportional—Achromatic Prisms and Lenses—Direct Vision Prisms—Anomalous Dispersion.

41. **The Spectrum.**—With the water prism as before described, or if the multiplying-glass or lustre-button used to produce the multiple deflected images has faces of small obliquity, nothing more may have been noticed than the refraction described in the last chapter; though even with these instruments attentive observation will generally discover a slight fringe of colour at the edges of the refracted images. We must now, however, employ a prism of more density and greater angle—either the glass prism (Fig. 9), or the prism bottle filled with bisulphide of carbon. Place in the optical slide-stage a perpendicular slit $\frac{3}{4}$ inch deep and $\frac{1}{8}$ inch wide, and arrange either the flint-glass prism, or the bisulphide prism on the stand, as in Fig. 49, first

focusing the slit upon the screen; and, as we expect the rays to be now very seriously deflected, turning the lantern off at a considerable angle before interposing the prism.[1] What a spectacle we have! There stands the glorious rainbow-band as first revealed to Newton's enraptured eyes; and which is to introduce us to a new and magnificent field—that of colour. Of all the people who have experimentally studied Optics, and who of course have performed this experiment scores and scores of times, never one yet but has felt that it never loses its fascination; the

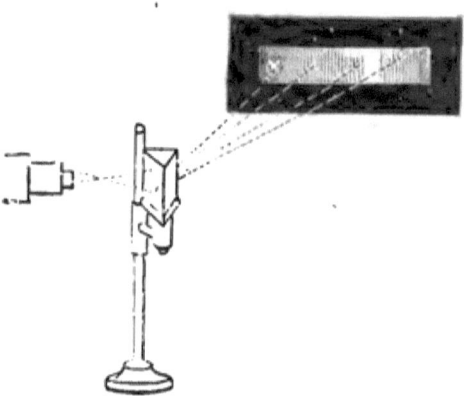

FIG. 49.—Production of the Spectrum.

same feeling of delight ever comes upon us, as that SPECTRUM appears on the screen (Plate II. A), which is to go with us, and be more or less, our guide, through great part of our future experiments. It is at once noticed, that while all the colours are bent aside, the red end of the spectrum is much less bent than the blue.

42. **Different Colours differently Refracted.**— Newton deduced from this and other experiments with the

[1] With a gas-burner the screen distance should not exceed about six feet.

spectrum, that each colour of light had its own degree of refrangibility, and that white light was compounded of various colours. We demonstrate the first point as follows. Arrange as before, but with a *short* as well as narrow slit in the optical stage—let it, say, be an aperture ⅛ inch square—and let its long *narrow* spectrum be projected by the bisulphide prism. We may perhaps think that the effect of the prism is merely of itself to spread or open the rays. To see if this is the case or not, we adjust behind the first prism, or between it and the screen, our glass prism in a *horizontal* position, with the refracting edge downward. If it be so, our spectrum will now be *generally* widened as well as refracted upwards. It is not, however ; the violet end is refracted up far more than the red end, and the spectrum appears on the screen askew, or slanting. The spectrum may be perhaps a little thickened or widened, but that is all (it will not be so if it is a good *long* or well dispersed one, such as is produced by employing two bisulphide prisms). Each colour, proceeding from the red end, simply appears more and more bent up. Hence the "dispersion," or opening out of the beam of white light into a spectrum of various colours, may be accounted for on the hypothesis of a different refrangibility for each colour, supposing only we find on experiment that such a combination of colours will compose white light.

43. **Each Colour has its own Angle of Total Reflection.**—Newton proved the special refrangibility of each colour by another still more beautiful experiment ; one of the most elegant ever devised. It depends on the facts already noticed, that the angle of total reflection must vary with the index of refraction (§ 32) ; the violet rays being totally reflected (because more refracted), at an angle which would allow the red rays to leave the denser medium. Newton therefore arranged an experiment as in Fig. 50, except that

he employed the parallel rays of the sun instead of those from the lime-light lantern. A perpendicular slit N is placed in the optical stage with the objective removed, or on the nozzle of the lantern if an adjustable slit is at command, (see Fig. 50, which shows all the arrangements in plan). As close to the slit as convenient, on a table-stand, simply

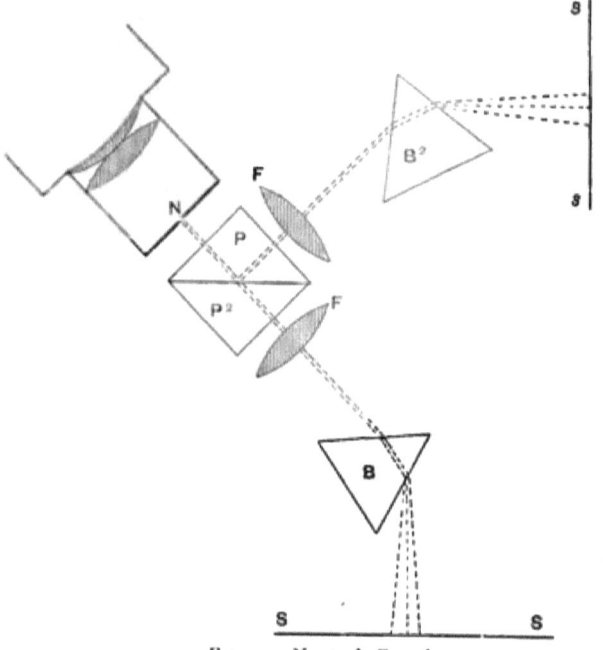

FIG. 50.—Newton's Experiment.

"stood up" on their ends, are two similar right-angled glass reflecting prisms, P and P2, with their reflecting sides together, kept together by an elastic band passed round near each end; they must not, however, quite touch, and may, if necessary, be kept apart by a narrow slip of paper at each end between them. In the direct path of the rays from the slit, is a focusing lens, F, and beyond that, on another table-stand, is placed a bisulphide prism-bottle, B, **in the** usual position for throwing a spectrum on the screen, s s. In the path of the rays totally reflected from the film of air between P and P2, is another focusing lens, F, and beyond that, on a third table-stand, a second bisulphide prism-bottle, B2, which throws its spectrum on the screen, *s s*, adjusted at right angles to the other screen. All being thus arranged in the general, the double prisms, P and P2, can be turned round their common perpendicular axis from right to left in the figure, till nearly all the rays from the slit pass through both, and the prism B throws a spectrum on the screen, s s, as usual. Except for a little loss by **reflection** and absorption, all is just as if P and P2 were not there, the *refraction* of one being exactly neutralised by the other, and the **rays** passing as if through one square bar of glass. Let now the double prism be **very** carefully and slowly turned round in the direction of **the hands** of a watch. At one certain point of revolution, just when the film of air meets the rays from the slit at the critical **angle** for violet, the violet *leaves the spectrum* on the screen s s, and, being totally reflected, appears on the screen *s s*.[1] Continuing very slowly to turn the double prism, all the colours in succession leave the spectrum on s s to appear simultaneously on **the** screen *s s*, so that if we letter the screens with the conventional names, at the point when one screen has only left on it the colours v o r, the other screen presents the missing colours, *v i b g*, which are "totally reflected."[2]

---

[1] The effect is less visible on this screen, enough light being always reflected from the air surface to give a little spectrum. But it can be seen that the violet is strengthened.

[2] Only the sun, or the small radiant point of the electric light, will give the phenomena *perfectly* in these details. A large gas-burner will not answer for the experiment at all. A "mixed" jet will perform it fairly if the condenser will throw nearly parallel rays. The "blow-through" form gives too large a radiant for a good parallel beam; but

This beautiful experiment, then, shows that each colour has, in addition to its own proper index of refraction for the same medium, and in consequence of it, its own proper angle of total reflection.

44. **Position of the Prism.**—Notice further that, as with the water-prism (§ 36), on turning the prism round its vertical axis (it may here be observed that the best place for a prism is nearly or a little beyond where the rays from the lantern cross; in this position all the light easily gets through it), a position is soon found in which the beam is least deflected. But still further observe that, on rotating the prism on its axis in one direction from this position, the ray is not only more refracted, but more dispersed: the spectrum is lengthened, particularly at the violet end. Rotating in the other direction, while the refraction also increases, the spectrum is shortened.

45. **Chromatic Aberrations.**—This fact has an important application. If the prism refracts the blue rays more than the red, then a lens must do the same, and will bring blue rays to a focus nearer the lens than the focus for the red rays. By placing in the slide-stage two apertures pretty close together in a black card, one covered with red gelatine and the other with blue, and focusing them on the screen with the large loose lens, it can readily be shown that this is the case; and thus, besides the "spherical aberration," already alluded to (§ 38), we have what is called "chromatic aberration," due to the fact that a single lens will not unite all colours accurately in the same focus.

46. **Correction of Aberrations.**—But we have here found a means whereby different *combinations* of even single or non-achromatic lenses may be arranged so as to

even with that, it is at least easily shown, by taking both prism-*bottles* away, and leaving the rest of the arrangement, that at a particular angle the direct image of the slit is reddish, and the reflected one bluish.

correct, to a considerable **extent,** both these aberrations. **In** detail, this must **of course be worked** out mathematically ; but a simple illustration of **a point** which often puzzles students is worth **while. We have** just found that the effects **of three prisms, A, B, C** (Fig. 51), of equal angles, variously inclined **to a** horizontal ray, coming, **say,** from the left hand, will **be** very different ; and therefore these **effects may** be made largely to counteract each other. But **these prisms** may, as before shown, be regarded as portions **of simple** plano-convex and bi-convex **lenses.** The **arrangement**

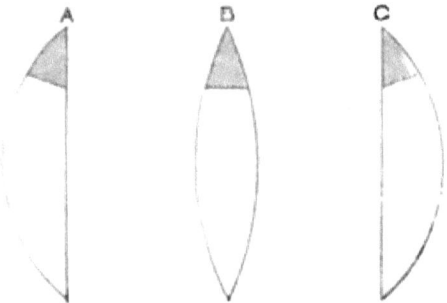

FIG. 51.—Effects of Position in Lenses.

adopted for our optical objective (Fig. 1) is **one** thus planned to correct a great deal of chromatic **and** spherical aberration by very simple means.

47. White Light compounded of the various Prismatic Colours.—Taking the different colours, as they **leave** our prism independently for the screen, or exist otherwise, we can show in many ways that when brought together they produce white. First of all, arrange a second prism B as in Fig. 52, so as to intercept the dispersed rays, and bend them back again, but leaving an interval of an inch or so between the prisms. We at' once restore the image of our slit on the screen, and it is *white*.

48. **Suppression of Colour produces Colour.**— Now interpose gradually **a black** card, **c,** between **the two** prisms, so as to suppress part **of the** spectrum produced by the **first** prism. We still get a sharp image of our slit, **and not** a spectrum; what are left **of the** coloured rays are accurately brought together again; **but it is** now a *coloured* image, and we learn an important lesson that holds good through nearly all future experiments. It is, that we almost invariably *get* **colour by,** in some way, taking away or *suppressing* colour. **If, instead of the** black card, we interpose **the edge of** a slightly wedge-shaped **prism, the** colour **taken away** appears

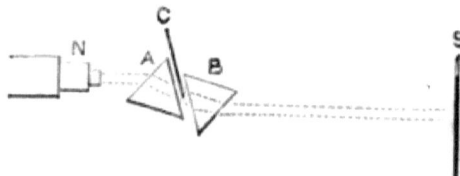

FIG. 52.—Recomposition of White Light.

separately, and is always "complementary" to the other — that is, together **they make white light.** (§ 71.)

49. **Experiments in Compounding Colours.**— A convex lens will also compress the colours together into a white image; and a cylindrical lens will do the same. The latter is rather expensive, **but** may be extemporised successfully by using a confectioner's glass jar 6 inches in diameter filled with **water.**[1] **Properly** adjusted between the prism and the **screen, this also will** compound out of the spectral rays a white, though not perhaps very sharp, image

---

[1] In **cold weather** the water should **be** slightly warmed, else condensation of moisture **upon the jar** will interpose tedious hindrances to **getting a good** image.

of the slit; and stopping off part of the spectrum will, as before, produce colour.

Another beautiful and striking method is shown in Figs. 53 and 54. Get seven bits of looking-glass ¾ inch wide by 2 inches long. From a round wooden rod an inch diameter cut discs, say ½ inch thick, as stands : to the top of each of these attach a bit of soft wax, and in this stick the end of a mirror, so as to stand vertically as Fig. 53. Arrange as shown in Fig. 54, standing the mirrors on a piece of blackened board, A, on a table-stand. First adjust the stand at such a distance that the rays from the prism P

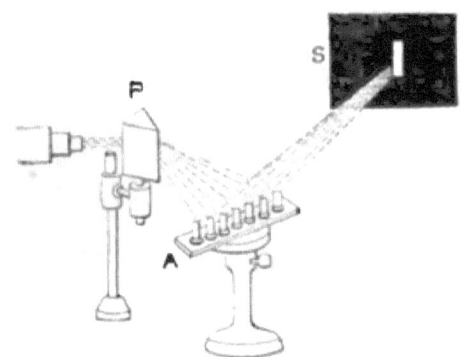

FIG. 53.   FIG. 54.—The Colours Recompounded.

about cover the breadth occupied by all the mirrors together, and then take off all but one at the end, and adjust that so that it may reflect its colour to a good spot on the screen, S. Put on the second, and turn that till its reflection occupies *the same spot;* so of the third and the rest. Note the changes of colour as we add colour after colour; till at last we have *white*. But take away—suppress—any one colour, and again we get colour.

Our next method depends on the persistence of visual impressions. We "see" things nearly a second after the exciting cause is gone—a long time, considering the apparently instantaneous character of all other light phenomena. Hold a slip of card in the diverging cone of light from the lantern; it is a mere strip of bright white. Cut rapidly through the cone as with a sword; it appears a white disc the size of the cone at the point of section.[1] Take, then, a white card circle, 12 inches diameter (Fig. 55), and divide into four quarters. On each of these paint in clear water-colours, as nearly as possible in the proportions of the spectrum, the seven spectral colours. You will not get them very correct at first, and it may be best to purchase the "Newton's disc," as it is called, of an optician;

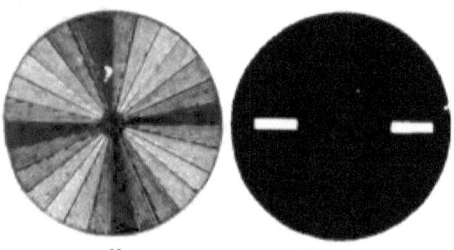

Fig. 55.   Fig. 56.

but many of such are too dark in the blue division. This is to be mounted so as to be rotated by a cord and simple multiplying arrangement, and then adjusted near the screen facing the lantern, so that all the light may be just about concentrated upon its face when the focusing lens is run fully out. If the disc of light is too large, place a circular aperture, cut in a black card, which will sufficiently

---

[1] I first saw this simple illustration employed by Professor Tyndall. Another familiar example is found in the circles of light produced by whirling round a lighted stick. Many toys of the "wheel of life" class depend on the same fact; and slides can be procured from any optician by which the same phenomena can be excellently shown with the lantern. They are known as Eidotrope slides.

reduce it, in the optical stage. This is important, for this fine experiment often fails because carried out by the general gas-light of the lecture-room, which gives very poor effect; whereas the full beam from the lantern brought on the face of the disc in a dark room appears very differently. Now rotate the disc rapidly, and we get *white* more or less bright or greyish, according to the correctness of the proportions. Newton's disc can also be purchased as a glass slide for the ordinary stage of the lantern, and shows very perfectly in this way, but it is inferior to the painted card for the next experiment. We see, either way, that the presentation of a proper assortment of the colours to the eye, *anyhow*, so as practically to mix them, produces white.

It is true, white produced in this latter way is, by comparison, greyish; and hence some of those who pride themselves upon being "practical" colourists, and despise the scientific investigations of those they term "theorisers," have denied that white can be really thus compounded. It was only very recently stated in print that the experiment only succeeds at all with "pale washes," and only produces a grey then; and this was stated with sufficient assurance to pass for positive knowledge. But it is simply due to ignorance of two points. In the first place, so far from requiring pale washes, the more vivid the colours are the better, if approximately pure; with some few discs I have seen a really very good white produced. And in the second place, a grey is the necessary and simple consequence of a *deficiency of light;* as can readily be proved by gradually diminishing the light thrown upon a really white disc of card—it gradually becomes grey. Now assuming that the spectrum may roughly be divided into seven colours, and that our white is produced by the successive presentation of these to the eye; a moment's thought will make it evident that, at the very most, only *one-seventh* of the light that a white disc would reflect, can be reflected by the coloured disc. In reality, **owing** to absorptions explained later on, it is very much **less**; and hence our white must be **more or** less of a grey if contrasted with a really **white** card.

We carry the last experiment a step further. **Cut** out from a circle **of** blackened paper, the size **of the** disc, radial sectors so arranged as **to** cover the same colours in each quadrant—say, the **violets and** blues—and fix them on the disc by a drawing-pin **at the** end of each. We thus suppress colour; and, **as before, we** get colour. Again, prepare **a** disc of stiff, blackened card, as in Fig. 56, with **two** radial slots each **2½** inches long, **and say,** ¼ inch wide. Run a drawing-pin **through the centre into the end of a** stick, so that it can **be rotated** by the finger **like** a child's windmill, and hold the affair, or fix the stick **in** the clamp, so that each **slot in** rotating will cross the nozzle of the lantern and so **let a** flash of light through. The nozzle itself should be covered with a cap in which **is a similar** slot, so as to make the flashes as nearly as possible instantaneous. (**This** rough-and-ready plan does just as well as more finished apparatus.) Let an assistant rotate the coloured disc, and **get a** good white. Now introduce the black disc and **rotate** that with the forefinger. **The** brief flashes prevent *all* the colours from now reaching the eye during one period of visual impression, and so again we virtually suppress more or less colour. Again, therefore, we get colour, its character and amount depending on the respective rates of rotation.

Yet another experiment, **on** the spectrum itself. Arrange **the** prism P as before in **front** of the nozzle, but with the **reflector** R so arranged that the dispersed rays from the slit fall on that, **and** are thence reflected to the screen, **as in** Fig. 57. The reflector must turn *very* easily in another vertical socket. **Give a** *rapid* rocking motion through a small arc to the reflector, which will, of course, move the spectrum backwards and forwards across the screen. Begin slowly, and you see that the motion in no way changes or modifies the colours. But when the motion becomes rapid, the spectrum becomes *white* over all the middle portion, colour being only visible towards the ends, where all the colours cannot, of necessity, be made to fall. The reason is the same as with Newton's disc, and, of course, if we stop off any of the coloured rays which fall on the mirror, by taking them away we again get colour.[1]

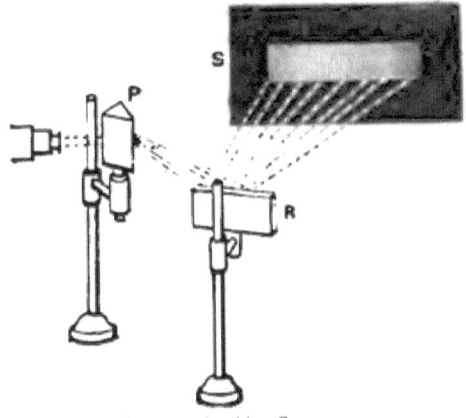

FIG. 57.—Rocking Spectrum.

50. **A Narrow Slit necessary for a Pure Spectrum.** —We can see from these experiments that to get a pure spectrum, or pure spectral colours, we must use a very narrow slit. For cut apertures in one of the black cards, as in B,C, Pl. II., the larger one being at least ¾ of an inch square. It will

---

[1] It is not every one who can, merely by hand, rock the mirror fast enough to get white in this way. By attaching a short arm to the mirror, linked to a small crank on a multiplying wheel, a rapid vibrating motion can easily be imparted mechanically. Such an apparatus is far superior to the mechanically-vibrated *prism* usually employed for such an experiment.

be obvious that the spectrum of this wide slit may be regarded as compounded out of the spectra of the narrow slits superposed; and from these it will be seen how **one** colour overlaps another. **If** the slit is wide enough, **a** glance demonstrates that towards the **centre** of its spectrum **all the** spectral colours, from some one or other of the narrow slits which **may** be regarded as its com**ponent parts, overlap** one another; **and so far as they do the middle** portion must be white, colour only **appearing at each end, where** the spectra of other portions **do not extend.** It is necessary to understand this, because it also has **been made a** difficulty by the " practical artist " school, and was **even brought** as an **objection** by Goethe against the theory **of the spectrum;** whereas a very little thought **over Plate II., B, C, will show that no other** result **was possible.**

The experiment also deserves pondering at some length, because it is a cardinal one. It shows that we may very possibly have colour really produced; but that if we use too large a body of light the visual effect may be masked, or hidden from us, by the *superpositions* of colours derived from various separate small portions which go to make up the whole. (Compare § 13.)

51. The Rainbow.—We are now in a position to deal experimentally with the beautiful phenomenon of the rainbow. We know this to be produced by the rays of the sun falling upon spherical drops of rain-water, and being thus refracted and reflected **back to** the eye; and if we fill a **glass** bulb with fluid, **and operate upon that to** represent on **a larger scale a single drop, we can trace** what happens. This beautiful experiment was first performed with a sunbeam, **by Antonio de Dominis,** Archbishop of Spalatro, about **1600 A.D.,** though Descartes seems to have usurped the **credit of** his investigations, as he also attempted to do with **the law of** sines discovered by Snell. We take a small

ARTIFICIAL RAINBOW.

glass bulb 1½ inch diameter, blown on a small tube, and fill it with water; or we may use filtered salt and water, for the reason that it not only increases the dispersion (§ 52), but diminishes the angle of deviation. The lantern must be turned *towards the spectator*, and placed further back; the objective removed, and a blackened cardboard or other cap with an aperture the size of the glass bulb, placed on the

FIG. 58.—Rainbow Experiment.

flange nozzle; round which, or on which, by a hole cut in the centre, is placed a screen, s, of white paper or cardboard. Instead of a card cap, the revolving diaphragm with various holes may of course be used. Adjust the lime-light to throw a *parallel* beam on the bulb B, between which and the spectators place, the blackened card screen (p. 17) to intercept the direct light. **The** bulb will **be** held in the clamp, c, by the stem, and **both** bulb and **fluid** must be brilliantly clear. **We** at once see **a** miniature **"rainbow"** reflected **back upon** the screen **round the lantern** nozzle, provided the **bulb** be not further from the nozzle than about the **radius of the** screen.[1]

It hardly need be said that **this rainbow is the real rainbow** *reversed;* any spot on the screen where red appears, means that an eye there would *see red* in the glass bulb; and each other colour, unless the **eye** was moved, would need another bulb in the proper relative position : but the experiment does show correctly the **emergence of a** nearly parallel chromatic bundle of rays at one **certain angle.**

That such **must be the case at one certain angle (about 40° for water) can be proved mathematically** from the **"law of sines,"** applied to the spherical bulb ; and the demonstration of this is really due to Descartes. He showed that at **one angle alone the rays, which at other angles of** incidence **emerged divergent and scattered, emerged as a** *nearly parallel* **beam, and thus produced conspicuous** phenomena **of** *some* **sort; the production of** colour by dispersion being afterwards explained by Newton's experiments with the prism. Fig. 59 shows in outline the course of the blue and red rays in both the inner and outer bow. The parallel rays, s s, from the sun falling on the drop *b* at the proper angle, are refracted twice and reflected once, so as to transmit red light to the eye ; and from the drop *a* (the angle *a* o *b* making about 2°), the blue rays being *more* refracted, also reach the eye ; *a b* **is therefore** the apparent breadth of the bow. Other solar rays, s′ s′, by *two* internal total **reflections** and two refractions, also transmit coloured rays, **but of course** fainter ; and these form the outer bow. An

---

[1] This experiment will not succeed with a plain gas-burner, a strong *parallel* beam being necessary.

inspection of the figure will show that the order of colours in the outer bow must be inverted. Theoretically several secondary bows are possible, and with a very bright sun three are occasionally seen; but as a rule only the primary and the first outer and inner secondary are sufficiently brilliant to be visible.

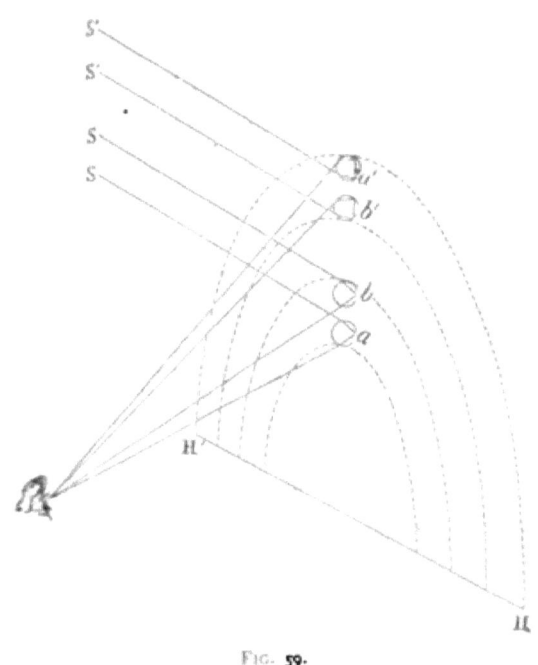

Fig. 59.

52. Refraction and Dispersion not Proportional.
—Make a water prism as in Fig. 60, by cementing with marine glue two slips of glass 6 **or** 7 inches long and 2 inches wide into a **V** trough with angle of 60°, with two partitions as well as two ends. If any difficulty about these, wooden partitions will do, cemented with black sealing-wax varnish. Fill one division with water, the next with saturated salt and water,

the third with saturated sugar of lead in water. Place the horizontal slit in the optical objective, or on the flange nozzle, and focus on the screen; then pass the three fluids in succession across the beam. Observe that the water refracts and disperses it somewhat; the brine more; the lead most of all. The natural and first conclusion would be that the two effects are always proportionate : that dispersion goes *with* refraction in due proportion. Newton so concluded, misled probably by the frequent use of lead in his water prisms, which masked the very low dispersive power of water. At all events, he made an experiment he thought decisive, immersing a glass prism *in* a water prism of variable

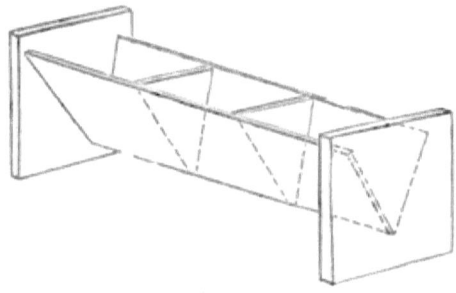

FIG. 60.

angle, with their refracting angles opposite. When the angle of the water prism was so adjusted that there was no deviation or refraction, he found no colour; and hence concluded that the dispersion of the glass and water were proportionate to their refractive powers. It is difficult to account for his result in any other way than that supposed ; for when Dollond repeated this very experiment with glass and water, the result was exactly the opposite : viz. when the refraction was exactly counteracted, colour or dispersion *remained ;* and when colour was nearly banished a considerable deviation remained—a discovery that led at once to the construction of achromatic lenses.

In fact, different media vary widely in the proportion of their dispersive and refractive powers. For its refracting or bending power, the dispersive (or spectrum-lengthening) power of bi-sulphide of carbon is enormous; and if a prism bottle be filled with a solution of the double iodide of mercury and potassium, as described by Dr. Lieving, prepared of the utmost density, while the blue end of the spectrum will be absorbed by the pale yellow solution, the green and red are dispersed nearly twice as much as by the bi-sulphide. Flint glass, again, refracts light little more than crown glass, but disperses it nearly twice as much for the same angle.

53. **Achromatic Lenses.**—Here, then, we have the power of correcting or destroying chromatic aberration. Dealing with prisms as the simplest case of the problem to be solved, a flint glass prism of little more than half the angle (in fact, the proportion depends entirely on the density of the flint) will counteract nearly all the dispersion of crown, but leave a considerable amount of refraction. Such a double prism can be bought for 5s. Focus a slit on the screen as before, and on a small table-stand place on end the crown glass prism. We get as usual refraction, and a spectrum; in this case rather a poor one, owing to the little dispersion of the crown glass. Now place next it, in contact, the flint prism, with its angle the reverse way: we still have the beam bent aside, but the *colour* is practically gone, and it is a white image of the slit, and not a spectrum, which appears on the screen. It is not necessary to explain in detail how this fact enables us to construct achromatic lenses.

54. **Direct Vision Prisms.**—Conversely, a prism of flint glass of about $52°$ (for average density), will counteract the whole refraction of a crown glass prism of $60°$, but will so much *more* than counteract its dispersion, that there will be

a considerable *reversed* spectrum. Hence we have the power of constructing "direct" prisms which give a spectrum without refracting the beam of light. Direct vision prisms composed of from three to five prisms of glass, are largely used in direct vision spectroscopes. For lantern work such prisms are very expensive, a large size being required; but Mr. C. D. Ahrens has introduced a prism as in Fig. 61, which answers our purposes at a very moderate price. G G are prisms of light glass, enclosing between them bi-sulphide of carbon, B. One made for me gives a dispersion more than equal to a prism-bottle of 60° without any deflection at all, and is not only very handy to work with, as obviating any turning of the lantern aside, but more light

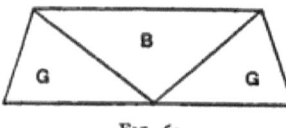

Fig. 61.

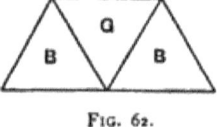

Fig. 62.

passes through, and the spectrum is not at all curved.[1] Just as these pages were written, Mr. Ahrens brought me for trial another bi-sulphide prism made as in Fig. 62. Here G is one equilateral prism of glass, projecting into the cell of fluid B B. There being only one ordinary glass prism here, this is the cheapest large compound prism that can be made. It gives enormous dispersion—about 60 per cent. more than that of a prism-bottle—with very little, if any, more than

---

[1] With a single prism there is a perceptible curvature in the transverse lines or edges of a spectrum projected on a screen, owing to the convergence, and consequent various angles, of the rays traversing it. To get rid of this curvature in spectroscopes is the object of the collimating lens. This brings to parallelism the diverging rays from the slit adjusted at its focus, which then traverse the prism at the same angle.

ordinary deflection, **and it** may probably be supplied **un**mounted for about 40s. Much more light traverses this prism than can get through the two prism-bottles generally used when great dispersion is required.

55. **Anomalous Dispersion.**—We have not even yet got to the **bottom of this** matter, however. We may obviously construct prisms **of** different substances, such as water or crown glass, and flint glass, of such angles that their respective spectra shall be of equal length. But if we do so we find the two spectra *do not agree*. There is more dispersion in one region than in another, as produced by one substance compared with the other; and hence *perfect* achromatism is impossible with only two prisms **or two lenses.**

But stranger phenomena still **await us.** There **is a** purplish-red aniline dye called Fuchsine. **Obtain a** little of it, and fill **a** prism-bottle with a *dilute* solution of the Fuchsine **in alcohol.** A bottle of the ordinary 60° angle can **be made to answer, but** if one **of 25° or** 30° can be procured it is rather **better.** There must also be provided a glass trough with parallel sides sufficiently wide apart to contain the prism-bottle; and this also, after the bottle is inserted, must be filled with the *same* alcohol that dissolves the Fuchsine. It will easily be understood that when the prism-bottle is immersed, both the refractive and dispersive powers of the *alcohol* in the bottle are exactly neutralized by that in the trough. This peculiar adjustment is necessary **to** obtain the Fuchsine action separately; for if we use the bottle of Fuchsine alone, supposing we fill it with a strong solution, the absorption is so strong that only the red rays get through (see § 69), while, if we dilute it, as we must do to get more **of** the spectral colours, any abnormal dispersion of the Fuchsine is overpowered by the ordinary dispersion of the alcohol. By exactly neutralizing the latter, then, we can employ **a** dilute solution, and still trace the effect by the lime light.[1] Success will now depend upon attaining by trial a proper strength, and passing the rays from the slit through the Fuchsine as close as possible to the refracting angle, or thinnest portion of coloured fluid. But when these matters are properly adjusted, it will be seen that the order of the spectrum is *totally altered*, we may almost say reversed. Counting from the red end, instead of beginning with red, the Fuchsine spectrum *begins with blue;* goes on to violet; then (after the absorption of some colours altogether, a matter we must study later on) comes red *in the middle*, ending with yellow; green being absent. How many of the colours can be discerned depends on the strength and thickness of solution.

Many other substances give similar phenomena, one of the best being cyan blue. Though startling, however, these appearances are not really more wonderful, when we attentively ponder them, than that dispersive power should, compared with refractive power, differ *at all* in various substances. Our experiment with the Fuchsine simply shows us in a more exaggerated and startling form, the very same fact—whatever it is—which makes the dispersion of flint differ from that of crown glass. All alike reveal the "anomalous dispersion" of light. We can no longer maintain that the colours have even an invariable *order* of refrangibility. This is generally the case; but we have now found that sometimes they have not.

And at this stage we must pause a moment to collect our

---

[1] For private observation only, two small slips of glass may be inclined at an angle of say 10° by a strip of wood placed between them at one edge, and a drop of *strong* Fuchsine solution placed between them. Through this prism a brilliantly-lighted slit may be observed from a good distance; and through such a small thickness even a solution strong enough to overpower the dispersion of the alcohol will allow sufficient of the colours to pass.

ideas. We accounted for reflection by a rough working hypothesis, which for years was more or less accepted, and was known as the Emission or Corpuscular Theory of Light. But even then we found some difficulties in it. These are now vastly increased in many ways; and we find ourselves once more, by the mental constitution bestowed upon us, bound to ask the question: What is Light? We must frame some intelligible hypothesis by which we may string together at least the foregoing facts, and which, if possible, may also account for such phenomena as we may yet further discover. This, then, will be the subject of the following chapter.

CHAPTER V.

WHAT IS LIGHT?—VELOCITY OF LIGHT.—THE UNDULATORY THEORY.

Light has a Velocity—Velocity implies Motion of some sort—The Emission Theory—Transmission or Motion of a State of Things—Transmission of Wave Motion—Illustrations—Wave-motion and Vision—Analysis of Wave Propagation—The Ether—Refraction according to the Wave Theory—Total Reflection, Dispersion, and Anomalous Dispersion—Mechanical Illustrations.

56. **Light has a Velocity.**—If a man strikes a bright light say ten miles off, we see it so instantaneously that it is very difficult at first not to believe that we are indeed, in some mysterious way, conscious of it the very instant it happens—that time has nothing to do with the matter. And that was probably the most ancient idea of the manner in which we "see" things.

But this idea was necessarily abandoned for ever after a discovery made by Roemer in 1676. He found that, taking the calculated time for the eclipses of Jupiter's satellites, they always took place eight minutes earlier when the earth was nearest to them, and eight minutes later when it was furthest away; whilst if the earth was at either of the mid-points, they happened at the average time. He very soon drew the unavoidable conclusion, that light required about a quarter of an hour to cross the earth's orbit. This was soon after confirmed by Bradley, who calculated the very same velocity independently, as nearly as could be, from the apparent "aberration" of the fixed stars.

At a later period the velocity was actually measured instrumentally, first by Fizeau and Foucault; later by Cornu; latest of all by Professors Young and Forbes in Scotland. The methods have been chiefly two. In Fizeau's method, a ray of light is sent between the teeth of a toothed wheel to a distance of a few miles; and thence reflected back in the same path. If the wheel is rotated swiftly, it is plain that while the ray has been journeying and returning

FIG. 63.—Fizeau's Experiment.

with a sufficient velocity, a tooth instead of a space will occupy its return path, and produce eclipse of the reflected ray. This effect must be gradual as the velocity is increased, as in Fig. 63, till total eclipse is produced. After this, if the velocity is further increased, the light will gradually reappear, to be again eclipsed; and the velocities of the wheels being known, the various eclipses will mutually check each other. Professors Young and Forbes introduced still more delicate checks, by transmitting the same ray to two different distances at the same time, by which they were able to detect the extraordinary fact that in air the velocity of red light appeared considerably less—perhaps one per cent. less—than that of blue.[1]

In Foucault's method, a **ray of** light, after **passing cross-wires to serve as** an image-point, proceeds any convenient distance to a **mirror** very swiftly rotated. It is thence reflected to a concave spherical mirror whose centre of curvature is the axis of rotation of the revolving mirror at the point struck **by the ray.** The **ray** is therefore reflected back **in** its own path to the **rotating mirror ;** and if this is at rest or rotated slowly, it is again reflected in **its** original path **of incidence, centrally upon the cross-wires. But** when the **velocity is sufficient, the mirror has** rotated **through a** small **angle while the ray has travelled to the concave mirror and back ;** and the return ray is therefore deflected to one side of the cross-wires, through an angle double that of the rotation of the **mirror** (§ **21).** This apparatus **is** so sensitive **as to** be applicable to **distances of** only **a few feet ;** and hence, by interposing a tube filled with water between the two mirrors, Foucault was able to prove absolutely that the velocity of light was considerably less in water than in air.

All these methods give the very same velocity for light

---

[1] This result is so startling that it must at present be received and reasoned about with much caution. According to the theory of refraction, **the blue rays should be, and are,** most retarded in air, though not to nearly so great an extent as **1** per cent. When the conclusion to which **Dr. Young and Professor Forbes had** come was first stated (at a meeting of the London Physical Society, **on June** 4, **1881),** Dr. Spottiswoode, Pres. R.S., Lord Rayleigh, and **others,** pointed out the difficulties it presented in accounting **for** the freedom from colour of Jupiter's satellites when emerging from the **planet.** Professor Forbes suggested that it might be accounted for **by** the gradual emergence : but it was **pointed out that at least** variable stars **ought** to appear coloured during rapid **increase of** their light. **At present** the conclusion is so momentous in its **consequences as to need the** most rigorous verification ; and it can only be **accounted** for on the supposition that the atmosphere is a remarkable **case of** " anomalous dispersion " (§ 55).

within quite a small percentage, much less than might have been expected in measuring such enormous velocities. The most recent determination of the **velocity by** Professors Young and Forbes is 187,200 miles per second.

57. **Light must be Motion.**—The velocity of light once proved, involves another point. Observe the absolute necessity **of** the case. **At** the moment one of Jupiter's satellites emerges from behind the planet, it sends out, somehow, a ray of light—whatever that may be. At a given moment that ray from the planet P has reached the nearest point A of the earth's orbit. A quarter of an hour later it has reached the furthest point B. In the interval, *something*

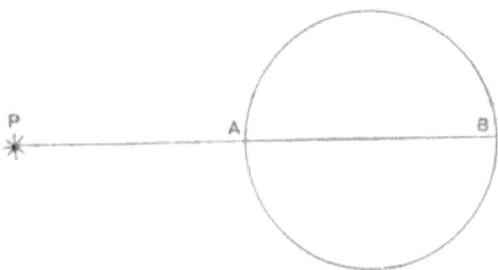

FIG. 64.—Light necessarily Motion.

or other has passed from A to B. That passage of something from one point to another is obviously motion; and the conclusion that Light is Motion *of some sort* is an absolute intellectual necessity: there is no possible escape from it. It only remains to discover what the motion consists of, or what it **is** that is moved.

58. **The Theory of Emission.**—The emission of very fine particles from the luminous body was the next and most natural idea, and it may be made to account for reflection, as we have seen; but it is encompassed even at that point with considerable difficulties. One of these we have considered already (§ 27); but it may be supposed the luminous

particles, to be seen, must actually enter the eye, and impinge upon the sensitive retina. But we have another and a greater difficulty, in the enormous *momentum* even the smallest particles must have, travelling at such enormous velocities as have been proved in the case of light. The weight of one grain would be equal in momentum to a large cannon ball as shot from the muzzle of a gun. Further still, what must be the *expulsive force* to produce such a velocity? And lastly, it is impossible to conceive of this expulsive force, and the velocity imparted, being the same for the largest and the smallest of all sorts of bodies, as is found to be the case. The feeblest candle-flame emits light at the same velocity as the enormous sun.

These difficulties might be partially evaded by supposing that the particles are so infinitely small, or otherwise so utterly different from common matter, as to have no relations of attraction with it; and, indeed, otherwise the propulsive or luminous body would, by its own attraction, gradually destroy the velocity, just as the earth gradually draws back a stone thrown up from it. But when Newton, who provisionally adopted this as a working hypothesis, though often leaning towards another, next to be considered, sought to account on such a basis for some rays being refracted while others were reflected, the only way he could do so was to suppose that some particles reached the reflecting surface in a repulsive "fit," others being attracted. He was confirmed in this by the "law of sines." Versed in the mathematical laws of forces, so that at times he almost thought in mathematical terms, his keen eye saw at once that this peculiar law was a *law of velocity;* and he suggested, therefore, that the refracted ray was dragged down or attracted by the glass or water, and had its velocity *augmented* in the proportion of the "index of refraction." It is odd (but rather general in algebra) that this proportion, equally with the reverse, accurately works out most optical problems. But it would follow that the particles *are* attracted by matter; and how, then, they can be projected with a velocity that never diminishes over stellar distances, is simply inconceivable. Finally, we have seen how it has been proved rigidly, that the velocity of light is *less* in the water than in the air. In absolutely proving this fact, Foucault and Fizeau absolutely overthrew the Emission Theory, and *that* idea of the "motion" of light. We have therefore to explain in some better way why light is *refracted ;* why at certain angles it cannot get out into a rarer medium, but is totally reflected ; (§ 32) and finally to account in some intelligible manner for the strange phenomena—totally unknown to Newton—of anomalous dispersion (§ 55).

59. **A State of Things may be Transmitted.**—Let us consider now another idea.[1] In Railway and Post Office we have a very familiar example of actual Things being sent from one place to another, or from a sender to a receiver, at a given measurable speed. That **may** answer in many respects to the old Emission Theory of Light ; and it is to be remembered that even in this case we must have some road or channel along which the sent Thing may pass, if it is not to be projected like a cannon ball. But let us think for a moment about a telegraph message. To quote Mr. Lockyer, here also "two instruments may be seen, one the *receiving* instrument, the other the *sender.* Between the office in which we may be, and the office with which communication is being made, there is a wire. We know that a Thing is not sent bodily along that wire, as the goods train carries the parcel. We have there, in fact, a condition of motion with which science at present is not absolutely

[1] For the whole idea of this paragraph, the simplest and most effective I have met with for illustrating what follows, I ought to acknowledge my indebtedness to Mr. J. Norman Lockyer, F.R.S.

familiar; but we picture what happens by supposing that we have a *state of things* which travels. The wire must be there to carry the message; and yet the wire does not carry the message in the same way as a train carries the parcel. It is further remarkable that the wire carries the *state of things* along, very much quicker than the train can move—in fact with a velocity commensurate with that of Light itself."

60. **Wave-motion and its Transmission.**—Let us now examine experimentally another kind of actual motion. Make a large groove in a piece of board, or a light trough like Fig. 65, about a yard long, in which bagatelle balls may roll. First roll one slowly along: we can measure the considerable time it takes to go from one end to the other. Now place in the trough a double set of 18 balls in contact; and drawing back one for some inches, roll it up to the rest at the same slow speed as before. Observe the difference. *Instantly*, to all appearance, the furthest ball now starts off as the first one strikes the row. It is not really so, for the time can be measured by proper appliances; but the eye cannot discern any interval.

Fig. 65.

Now *this is wave-motion*, of one sort. The first ball struck is compressed, and then expands, so passing on the compression; and thus to the last. Each ball executes or has impressed on it *small movements exactly like that of its preceding neighbour, but a very little later in time*. That is the entire essence of wave-motion; which may be of many forms, but with this property common to them all.

In the lantern we may illustrate this as in Fig. 66. Between two glass plates the size of the ordinary lantern slides, are cemented pieces of wood shaped to gentle curves as in the figure, of a thickness to just allow small balls of glass or ivory (the smallest procurable) to roll easily in the channels for them. In one, let a row of these balls rest, leaving the other clear. From the top of each slope let roll one of the balls. The one will be seen travelling across the screen: the motion in the other case will appear to be transferred *instantaneously* to the last ball in the series.

In a pond of still water, or in the middle of a circular sponge-bath filled with water, drop some small body, and study the beautiful phenomena. The spot into which the object fell is surrounded by a circular *wave* of water, which travels quickly outwards. This apparent motion is again

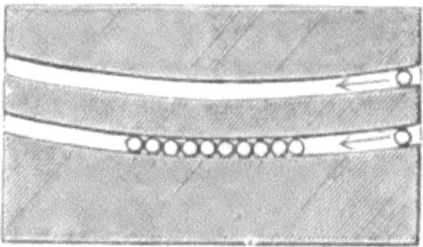

FIG. 66.—Slide for Rolling Balls.

very much quicker than any actual motion we could impart to the mass of water itself. It looks, too, as if the water were really proceeding outwards—being poured out as it were—from the central point. But if we drop one or two bits of paper on the water, we find it is not so: the paper only moves up and down, with a very slight horizontal motion indeed, which also is not continuously outwards, but *to and fro*. In reality the paper at any point moves in an elliptic orbit; but each particle moves in a similar orbit to its predecessor, a little later in time. Any one particle does not travel outwards, but *sets in similar motion the next particle*, and thus **the *state of things*** we call the wave **is** propagated.

It can be seen immediately that if the motion **of** light be anything of **this** kind, we have got rid at once **of** our greatest difficulty, that of the enormous velocity; for **wave**-motion is **always** much quicker than actual translatory motion. Across a deep open sea like the Indian Ocean, the great tidal wave **moves at the rate of** 1,000 miles per hour, though the **water** itself **could move** at nothing like it. In fact, if **we can only find the** idea otherwise fits the phenomena, we have got rid of nearly all our difficulties at a step. We therefore must trace the matter further.

If the balls just now referred to **were tied together** by **elastic** threads, we should have a *return* **wave from the last ball, since** it would **be** sharply pulled back **in its effort to escape.** We have all **seen this in** an engine run up against a train, when each carriage or **truck** vibrates several **times to** and fro, giving us a picture **of such** waves in any *elastic* medium, such as sound-waves **in air.** These give **us** still another form **of wave. In this case** every particle **of air** moves **to and fro in the** direction of the **wave,** but otherwise keeps **its own place, and** still the wave only, and not any particle **of air bodily, moves onward.** This fundamental **property which characterises all true** wave-motion is difficult to picture **clearly in** the mind : **but it is** so important to do **so,** that we again have recourse to our lantern, which will place it **before our very** eyes. **Get a circular glass** plate, **13 or 14** inches diameter, mounted on an axle which can be turned. **Exactly in the centre** of the axle-hole describe **a** circle ¼ inch diameter, and divide it into 12 portions, as in Fig. 67. **(It is near enough to** divide off into six by the radius, and bisect each division by the eye.) Arrange all firmly on a table, and having blackened the disc **all over,** take a foot-rule divided into eighths **as a** standard, and with a 3-inch radius strike or scratch a circle from say the top division. Extend the compasses exactly $\frac{1}{8}$th inch and strike another *from the next division;* then another $\frac{1}{8}$th inch, and so from the next division in the same direction, and so on, till we have gone twice round the divisions and struck 24 circles, giving us a band of scratches 3 inches across, with a little margin outside. (See Fig. 68.) Remove the objective and place on the flange-nozzle of the lantern, a cap with a fine horizontal slit A B, 3 inches long, or a glass cap

FIG. 67.

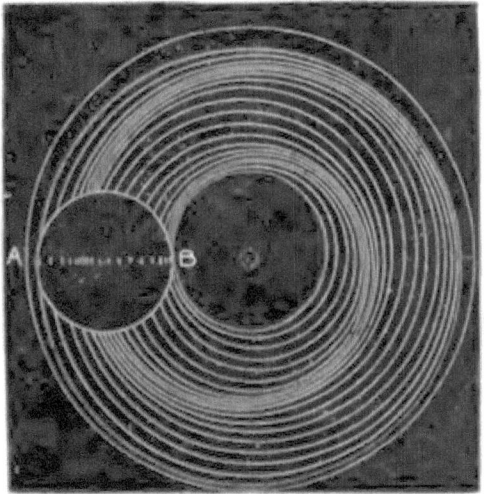

FIG. 68.—Crova's Disc.

with a line scratched in black; and arrange the disc [1] in front of this, so that the band of circles crosses the slit as in Fig. 68. In front of all adjust the loose focusing lens, and focus on the screen. On now revolving the disc we see a "wave" of alternate compression and expansion beautifully

[1] It is known as Crova's wave-motion disc.

delineated in actual motion : and by keeping the eye on any one bright dot, the precise nature of the motion will be understood better than by all the description in the world.[1]

For reasons future experiments will make clear, the vibrations which constitute light are, however, to be considered as like those of water-waves, *i.e. across* the path of the wave, not in it. A disc with a wave-line round it like Fig. 69, will give a moving image of this if revolved across a blackened glass cap scratched with perpendicular lines

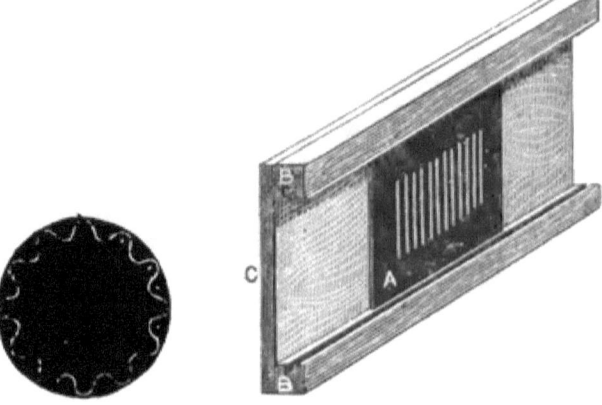

Fig. 69.   Fig. 70.—Wave Slide.

all over; but a better way is to make a "slide" as in Fig. 70, consisting of a fixed blackened glass A, in front of which are open grooves B B, the whole length, for a panoramic glass, 18 inches long, to slide freely. The slide is kept together by a slab of wood C, in which is an aperture the size of the field. On the fixed glass scratch fine perpendicular lines $\frac{1}{16}$ of an inch apart. On the long glass,

[1] The private student may draw the lines pretty thickly in black on a white card, and revolve the card under a slit cut in another card.

also blackened, a single wave-line may be scratched; but in order to show the interferences of waves at a later stage, it is better for the sliding-glass to be in three widths, kept edge to edge by a thin brass binding at each end, and with *four* sets of waves as in Fig. 71, one pair twice as long as the other. The centre strip has a wave of each length, and by drawing it along through the binding the length of the short wave, as drawn below, it can be shown how the same given retardation brings the short wave a whole vibration, and the long one half a vibration behind their fellow waves. It will be evident that when the whole arrangement is placed in the ordinary lantern-slide stage and focused, and the

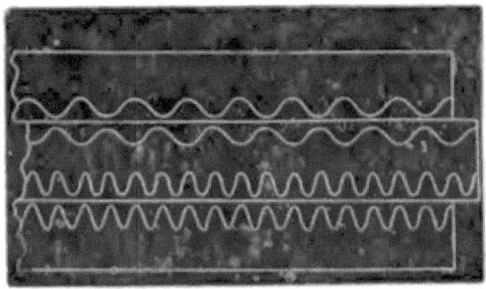

FIG. 71.—Movable part of Wave Slide.

sliding part moved along, we shall have *moving* waves of white spots.[1] Now it is to be again carefully noticed that *no spot advances at all.* Every spot simply moves *up and down*, while the *wave-form* only advances across the screen, being due, as before, to the fact that each spot, representing an atom of ether, moves a *little later* than its predecessor in a similar path.

Thus we see that, by this kind of mechanism (actually

[1] This slide was the result of much consideration; and after trying both, I consider it much superior in effect to the disc usually employed, besides being much more easily made and adjusted.

proved to exist in the case of the ivory balls, sound, and countless other motions), motion imparted at one end of the line of transmission is *yielded up* with enormous rapidity at the other, not only without any inconceivable motion between, but whilst things generally between, as in the case of our ivory balls, may almost seem to be in a state of rest.

61. **Wave Motion and Vision.**—It is not difficult to conceive how such wave-motion as that shown by our slide may affect the eye. We can easily imagine how longitudinal pulses of air beat on the drum of the ear D, as in Fig. 72; and if we imagine the retina to be furnished, as at R, with perpendicular rods standing up like a cat's whiskers, we can readily conceive how the transverse motion of the last

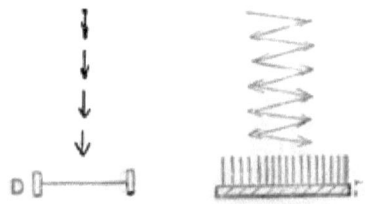

FIG. 72.—Sound Waves and Light Waves.

particles in the wave should excite these. Those who are so happy as to possess whiskers, can easily prove by experiment that a brush across them yields more sensation than perpendicular pressure. Now it is remarkable that late microscopical researches do show the retina to be studded with fine rods; and though the matter is yet under investigation, it appears probable that this is the true method of vision, or at least, that by which the optical image on the retina is transformed to consciousness in the brain, upon which our "seeing" Light *as* Light obviously depends. Several future experiments will lend much strength to this view; and another weighty confirmation of it is found in

v.] ANALYSIS OF WAVES. 101

the well-known fact that pressure on the eyeball produces the sensation of light.

62. Analysis of Wave Propagation.—We have not solved all the problem, however. If light consisted of particles

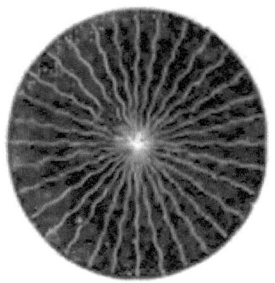

FIG. 73.

emitted, or if even wave-motion were only propagated in radial lines from each centre (and this is how some people quite erroneously understand the Undulatory Theory) it will be obvious that as the distance becomes greater, the radii must separate as in Fig. 73, and we should have spaces

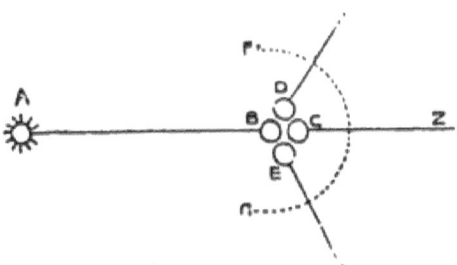

FIG. 74.—New Centres of Wave-Motion.

between with no wave-motion. Yet we see an unbroken circle of waves in our pond; how is this? Figs. 74 and 75 supply the answer. In Fig. 74 let us consider a wave direction, A B, from the luminous centre A, and suppose it arrived at the particle B. The next particle to receive the motion *in the same line* will be C. But C only receives it because it is the *next* particle; and obviously the particles D and E are just as near, and in the very same kind of contact; *they* ought also to receive the motion. They do, and the consequence is that every vibrating atom over the whole sphere surrounding the wave-centre becomes a new centre for another sphere F G. Thus in Fig. 75, let us consider the waves from the centre L to have reached the circle M N. Every *point* in this circle sends out fresh circles of waves, shown on the curve O P. But analysis shows that

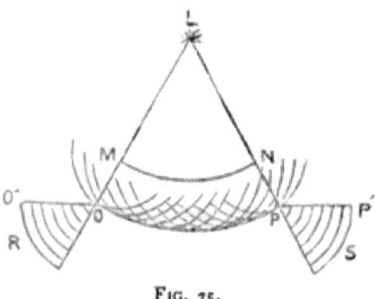

FIG. 75.

these destroy each other by mutual interference (see Chap. IX.) except in the grand circle O P, where they reinforce each other; and thus the curve O P remains as in a pond, the grand front of a main wave. If, however, we interpose a screen, at O O' P P', which stops this main wave, then the subsidiary waves which go to make it up are only partially destroyed by interference, and appear at R S, as in the diffraction fringes hereafter to be seen (§ 106.) Most of the difficulties found in comprehending the wave-theory are entirely owing to not understanding this "construction" of the waves; and it is hardly too much to say that while, on the one hand, any apparent objection that can be brought against the Undulatory **Theory** applies **with** equal or **more** force to *any other* theory ; **on** the other hand, any rays, **when** isolated and studied by themselves in detail, behave exactly as the theory would lead us to expect. The correct analysis of these details of wave-propagation **is originally** due to Huyghens.

63. **The Ether.**—If, however, waves are sent from distant stars **to our eyes,** and elsewhere, there must, as in all the preceding cases, be some medium through which they are propagated. We know by experiment on other wave-propagations, such as those of sound, the general properties such **a** medium must have ; since **it is** known that the velocity of sound is directly proportional to **the** square root **of** the elasticity of a **medium, and** inversely **as** the square root of its **density.** This **has been proved by** rigid experiments in many gases ; so that, for **instance, the** velocity in hydrogen, a **rare** gas, is much greater than in air. The *enormous* velocity of light-waves, therefore, is only possible in a medium **which is** almost infinitely elastic, **and at** the same time almost infinitely rare, and **yet** which **has a** *definite* proportion between these two functions. **It cannot** be attenuated air, but must be something far more rare ; since **we** can keep air out of a glass vessel, while light passes through glass and many bodies, and through the best vacuum we can form ; moreover we know that air absorbs or stops light considerably, and hence light could never reach us, through the rarest atmosphere, from the enormous distances of **the** stars. Still further, attenuated air could not give us those *definite* **velocities** with which we have **to deal,** but **must give a** widely different one from that in denser **air ; nor, finally, is** the proportion of its elasticity to its density anything like great enough. We have therefore to conceive of some still more rare and subtle medium which fills all space, and which is called the Ether. We

shall hereafter see that we are obliged to attribute to it other physical properties quite different from those of an atmosphere.[1]

But though this ether easily permeates all we call matter, it is perfectly easy to conceive of its particles, and those of matter, hindering and otherwise acting on each other, or communicating motion to each other. Wind passes freely through a hedge of trees; but still the trees hinder or "slow" it, while yet again the wind moves the trees; and conversely, if we could shake the trees, we should cause a wind. Such, in brief, is the conception or hypothesis of the ether as held by physicists.

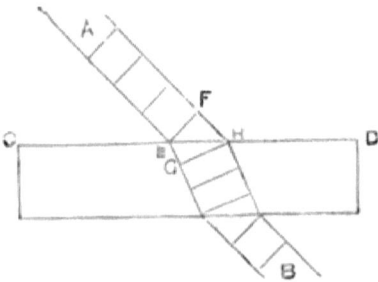

FIG. 76.—Refraction.

64. Explanation of Refraction.—Refraction now is easily explained. Any beam of light has a *wave-front across it*, and it is obvious that in meeting any refracting surface obliquely, one part of this wave-front will meet it before another. Conceive, then, that while the ether permeates the open structure of all matter, it is still *hindered* in its motions by it, as the wind is hindered, but not stopped, by the trees. Then trace a ray A B (Fig. 76) to the refracting

[1] Some attempts have lately been made to dispute the existence of an ether (see *Phil. Mag.* April, 1879, and July, 1881). It appears to me that the objectors have not borne in mind the essential physical necessities of the case.

surface C D, marking off the assumed length of its waves by the transverse lines. The **front** will be *retarded* at E before it is retarded at **F,** and we may assume the retardation is such that the wave in the denser medium is **only** propagated to **G, while** in the rarer medium it reaches **H.** It is plain **the beam must** swing round ; but when the side F also reaches the denser medium, the whole will be **retarded** alike and **the beam** will proceed as before, only slower and in **a** different direction. The theory exactly fits **all the** phenomena ; and when we come to polarisation it will **be easily** seen why the beam is *divided*, and part reflected while part is refracted (§ 119.)

We may illustrate this comparative retardation **in a** comparatively denser medium by a very simple but **beautiful** experiment. Take a large thin glass, or **any** other kind of sonorous "**bell**" **upon a foot** (Fig. 77), **and** drawing **a** violin-bow across the **edge, we** easily **produce a** musical sound. By adjusting **a** small **wire point to** barely touch it, as on the left hand, **or hanging a small** pith-ball barely against it, **as on the** right, we easily **prove that the sound** is due **to** rapid **vibrations or** motions, **which** at a certain speed produce a given note. Fill the glass with water, within say an inch of the top, and it gives a much graver note, showing, as we know, much *slower vibrations.* It might be supposed **that** this is merely due to the added weight of the water *in* **the** glass, **as** we lower a tuning-fork by loading it, the water vibrating with the glass as one whole. But it is not so ; for if **we** empty the glass, and immerse **it** in some flat-bottomed vessel which, when filled to the brim, brings the water outside the glass precisely to the same level as before, still leaving **the** glass an inch above, so that it can be excited by **the** bow, **we get** the very same lowered note. Finally, if **we now** fill the glass as well, the note is still further lowered ; showing us that the motion is simply more *retarded* in the denser medium, though the glass can still move freely in it.

65. **Total Reflection Explained.**—Total reflection can be explained by mathematical analysis, and is so explained by Airy, while it is impossible to explain it by any

FIG. 77.—Vibrations in a Glass.

other theory that has ever yet been framed. Airy shows that, beyond a certain angle, no *main* wave can emerge, while the small secondary waves perish immediately by interference (Chap. IX.).

66. **Dispersion Explained.**—The different refrangibility of different colours is easily accounted for. According to both Emission and Undulatory theories the measurements agree, and those who believed the Emission theory, equally had to consider the red particles as larger or stronger than violet ones, as we have to consider violet waves only half the length of red ones. But as all colours of light seem to move with equal velocity in ether, violet waves must make *two* vibrations for each one of red. If, then, the vibrations are retarded at all in a refracting medium, on the face of things those which occur twice as often will as a rule be hindered most: they have more of the hindered work to do, and, therefore, must be more refracted, provided there be nothing so peculiar in the arrangements of the molecules which retard them, as to affect them otherwise.

67. **Anomalous Dispersion Explained.**—But there may be relations in the retarding molecules of matter which do affect the ether-motions peculiarly, and so cause " anomalous " dispersion. We cannot suppose that the *lengths* of the waves for each colour preserve under all circumstances a uniform proportion; for we have not only seen they do not, but our very theory of refraction itself, is based on the supposition that the wave-lengths are *shortened* by retardation in a refracting medium. This is assumed to occur, as we know, because any given particle of ether is more or less hindered by particles of matter from communicating its motion to the next particles. The only supposition possible is, therefore, that the *number* of vibrations per second, or their *time period*, is the determining constituent of colour, as it is of musical notes. Now we shall find strong presumptive proofs hereafter (§ 79) that the molecules of different bodies, like small pendulums of given lengths, have in fact their fixed and proper time-periods of vibration. And we can thus very easily conceive how certain relations

of these respective periods to each other, or of the lengths of the waves to the distances between the molecules, should cause molecules of matter to hinder or assist in very various degrees, the differing periods of vibration in the ether-waves. Cauchy has even shown mathematically, that dispersion at all, or the fact of different colours being differently refracted, can only be explained if the distance from molecule to molecule be fairly comparable with the length of the waves. (See § 111.)

68. **Mechanical Illustrations.**—We may illustrate practically many of these points by admirable experimental means, due to the ingenuity of Mr. Tylor.[1] Though not strictly "lantern" experiments, they are easily seen by a

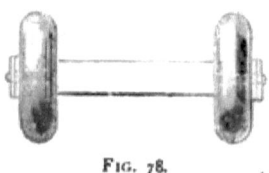

FIG. 78.

whole class, and vividly illustrate the subject. We are dealing with *motions*, supposed to be in free ether *equal* at any two points of a wave-front exactly transverse to the direction of the ray. It is obvious this is a purely mechanical problem, and that we may mechanically represent it with fair accuracy by two equal wheels revolving freely on two ends of an axle, and left to roll down a slightly inclined board. Let Fig. 78 represent such a pair of wheels, the axle being made of $\frac{1}{2}$ inch iron, turned down at the ends to about $\frac{1}{8}$ inch, on which revolve freely but accurately boxwood wheels about 2 inches diameter, with rounded rims. These dimensions were found best by Mr. Tylor.

[1] See *Nature*, Jan. 1, 1874.

It is obvious that when **rolled** along **a** smooth board such an axle will preserve one direction. **But** referring now to Fig. 79, let there be glued on the board a rectangular and a triangular piece **of** the thick pile plush called "imitation sealskin." If this is presented to the wheel-track the right

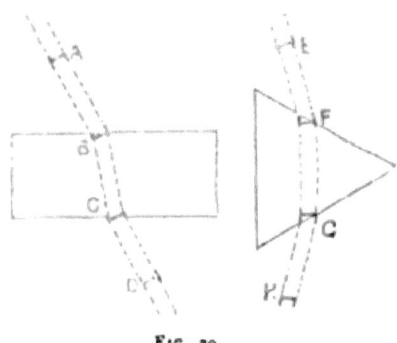

Fig. 79.

way of the pile, **it** will **retard the** motion considerably, and when the track A B is oblique, the wheel that first meets the plush **being first** hindered, the track will swing round to the

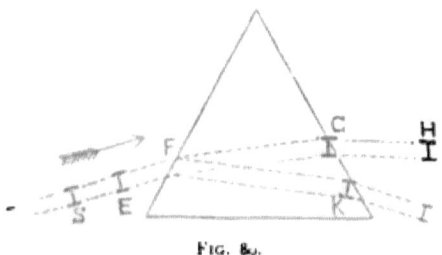

Fig. 80.

direction B **c, and on** leaving the plush resume the track C D, exactly picturing the course of a ray of light through a thick piece of glass. The triangular piece in the same way represents a prism, the track E F being refracted to F G, and thence to G H. Nay, even *dispersion* may be thus pictured by having a second pair of wheels, s (Fig. 80), of smaller diameter, to represent smaller waves—say 1¼ inches to 1½ inches. These wheels will be found perceptibly *more* deflected from s f to the track f k. The wheels may either roll freely down a slight incline, or may be held back by a thread at the centre of the axle. Finally, *total reflection* may be illustrated as in Fig. 81, for it will be found that

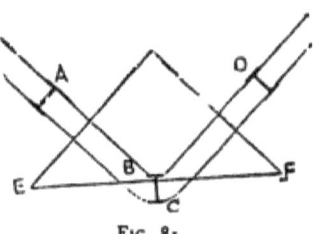

Fig. 81.

if the track A B leaves the edge E F of the velvet at a certain angle, the wheel c, which first emerges, gains so much on the one still upon the velvet, that the axle swings right round and proceeds on the track B D.

These experimental illustrations will sufficiently enable us to grasp vividly all the main points of the wave theory. We shall now resume the experimental study of colour.

CHAPTER VI.

COLOUR.

Absorption of Colours—What it means—Absorbed, Reflected, and Transmitted Colours—Complementary Colours—The Eye **cannot** judge of Colour-waves—Mixtures of Lights and Pigments, **and their** Difference—Primary Colour Sensations—Not the same thing **as** Primary Colours—Colour as we see it **only** a Sensation—Experiments showing merely sensational Colour.

WE have now cleared the way for another class of experiments, for which, to work **with comfort, we** must somewhat alter our arrangements by removing the objective, and placing **on** the flange-nozzle of the lantern a black card **or** other cap with a perpendicular slit cut in it rather longer than we have hitherto worked with—say $\frac{1}{8}$ inch to $\frac{3}{16}$ inch wide by $1\frac{1}{2}$ inches long.[1] A long slit, **owing to the convergence** of rays by the lens, gives a perceptible **curvature across the** spectrum band; but this need not matter **to us.** Arrange the loose focusing lens F (the longest-focus **one** if there are two) so as to focus the slit on the screen if the beam were direct (the lantern must, of course, be deflected, as for all prism work), and adjust the prism P otherwise as before. The whole arrangement is shown in Fig. 82, and its object is simply to produce on the screen the spectrum

---

[1] A brass cap with adjustable slit is, of course, much more convenient.

of a slit upon which we can more readily make various experiments.

69. **Absorption of Colours**—Providing now some coloured glasses, or some strips of coloured gelatine between glass plates, we make some experiments which teach us a very important lesson. We are apt to think that the sunlight which comes through a red glass window is all *turned into* red—made red. Well, there is the spectrum of our complete or white light on the screen, drawn out into its constituent colours. Over half the slit hold a bit of the red

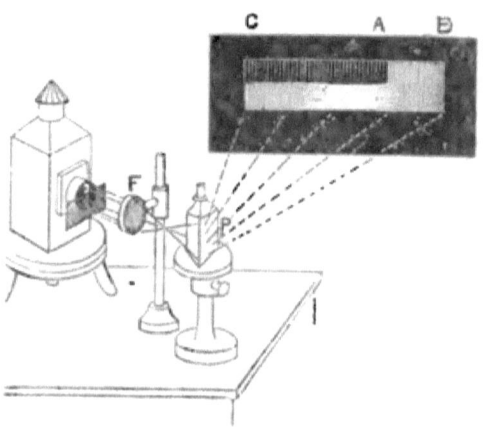

FIG. 82.—Spectrum Work.

glass; if the light, or most of it, is really "reddened," all the spectrum ought to be turned into red. It is no such thing, however. There is no colour in the spectrum of the glass *where that colour does not exist in the ordinary spectrum;* the sole effect is that certain colours are cut out, or absent. We get the colour, as so often before, by *suppressing* colour. If the glass is a pretty pure red, only red and a little orange, A B, is seen in the spectrum of the half slit covered by the glass; all the rest is cut away. So of all the other gelatines or glasses, but we soon find it is very difficult to find a *pure* colour; generally there are left, at least, two well-marked colours; and if we unite just those portions of the ordinary spectrum, by employing proper slits in proper places, and uniting the colour passing through them by our confectioner's jar, or a cylindrical lens (§ 49) we get the same colour as the coloured glass.

We see, therefore, that in passing through a transparent body, its molecules take up or absorb the waves of certain periods, and the remainder passing through give the colour of the glass. This is not at all difficult to understand. We have supposed every matter-molecule to have its own period of vibration (see next chapter); or perhaps more often periods, as it seems probable most molecules are complex. These molecules can freely communicate any vibrations to the ether-atoms; but conversely it must be different: the matter-molecules can only take up *synchronous* vibrations. That one tuning-fork will communicate its vibrations to another of **the** same note we know; and we also find that in thus giving up its motion to the second, it loses its own more quickly than one that does not. A fork mounted on a unisonal resonance-case sounds louder, but *stops sooner*, than one unmounted: it has been imparting its motion to the unisonal column of air, and in so doing exhausts its energy. See then what must happen. If waves which produce red sensations require vibrations of 450 million millions of times in a second, and such rays pass with others through matter whose molecules vibrate at that rate if set in motion, these must themselves take up or absorb all such from the ether, and the rest passing through give a colour due to the other rays—in this case green.

We shall examine tests of this theory presently; but meantime, however absorption is produced, many other experiments will show that the colour of our glass is produced simply by the absorption of certain colours. Throw a good long spectrum on the screen—through two prisms if the light is powerful enough—and provide some largish squares of the coloured glass or gelatine. Say we have a red one. Walk up nearish the screen and hold the square in the hand by one corner in the red rays; it stops these very little, or is "transparent" to these rays, which it permits to reach the screen. Move the glass along in the spectrum, however, and gradually we find it casts more and more shadow, till at last the shadow is *black;* the glass is absolutely opaque to such colours of the spectrum; it absorbs them all.

We soon find also, that as absorption increases, not only the shade, but what we call the "colour" itself, often changes, more and more of the spectrum being cut out; as indeed we should expect. With marine glue cement together a few glass troughs, of plates about 4 inches square and $\frac{1}{2}$ inch apart. Make solutions of chlorophyll (green leaves in alcohol, pretty fresh), permanganate of potash, picric acid, ammoniated sulphate of copper, oxide of copper in ammonia, and bichromate of potash. Try first the chlorophyll. One cell of moderate strength cuts out all the blue half and some bands in the red end; add another cell of the same, and the light transmitted is probably only the extreme red. A very few experiments of this kind with the other solutions will show how wonderful are the powers of this "spectrum analysis," for such it is, and how complex are most of the colours of natural bodies; and will prepare us for the further details of the next chapter.

70. **Absorbed, Reflected, and Transmitted Colours.**—But meantime a further lesson of the same sort. We should expect from the foregoing that colour transmitted through bodies, would often differ considerably from colour reflected by them. A red glass, since

it absorbs the green **and blue and** transmits the red, reflects hardly any, **and** appears by reflected light almost black, as do some other colours. In some cases the colours reflected and transmitted **are nearly** complementary (§ 71), but seldom quite so; because **even** reflected **light** has penetrated some distance among the molecules of a body, and thus been partially subject to their absorbing influence, or had certain vibrations taken up by them. Two very simple and pretty examples will sufficiently illustrate this. Place a single film of gold leaf smoothly between two glasses 2¼ inches by 4 inches, and bind round the edges to save from injury; or place it between two discs of glass mounted with putty in wooden frames that size. Deflect the lantern at right angles with the line to the **centre of** screen, throw the light on the gold leaf [1] **at 45°, and focus** it with the **loose** lens on the screen; **we get the yellow** reflected image. Now place it in the optical stage, direct the lantern to the screen, and focus: we get a transmitted *green* image. Take, **again, a clear glass, and** another blackened **at** the back; and cover each **with** a film of Judd and Co.'s red ink, or any other which owes its colour to aniline **dye. When** dry, place in the optical stage the clear glass: it transmits a fine *red* image. The blackened one reflects (when treated like **the** gold leaf just now) a beautiful *yellowish green* image. And **the** clear glass illustrates the usually compound nature (*i.e.* part transmitted through a portion of the substance and so partially absorbed, and part reflected) of reflected light, by giving a *reddish orange* image.

We can easily prove by experiment that the colours we see in natural objects are chiefly residuals left after this internal

---

[1] In focusing such surfaces by reflected light, all the light from the nozzle should be concentrated on the surface, and some small mark, such as a black dot, sharply focused on the screen by the loose lens.

absorption, or are colours to which the bodies are transparent. Get **any flower which** shows a full green leaf, and rich red petals in **largish masses,** such as a tulip, **and we soon find we cannot** " see " its colours unless they **are** either placed **in** white **light,** or in the appropriate colours **of the** spectrum. **Throw** once more on **the** screen the prismatic **band, and** move along the tulip in its rays. In the red rays the red flower shines bright red, and the leaves possibly dull red (owing to the peculiar spectrum of chlorophyll, which transmits the extreme red **as well** as the green). **But as** we move **it along the** red **becomes** black, and the **green changes also, precisely as** the spectrum **did when we cast upon it the shadows of** our coloured glasses. **Coloured flannels or highly-coloured** chromo-lithographs, **moved along in the spectrum, also** yield very instructive phenomena **of the same kind.**

71. **Complementary Colours.**—**We** have found that **we** make **white** light by compounding together all the colours of the spectrum ; but we have also found that we may produce white with much less ; for in the experiment with seven little **mirrors (§ 49),** many of the prismatic rays were **of** necessity omitted. **We** carry this **method** of experiment **further, and** arrange our prism **as before,** with either a **cylindrical lens (or the** water-jar **as** such), or a large lens properly **adjusted, to re-form a white** image **on** the screen. Now prepare two **black** cards with slits $\frac{1}{8}$ inch wide and an inch long, **and insert them, as in Fig. 83,** in a strip of **blackened wood, with a** saw-cut in it, so that we can slide **and** adjust the slits **at variable** distances. Introduce this **between the** prism and re-uniting apparatus, and, first **covering up** one slit, **arrange** the other so that only the blue **rays pass** through **it,** giving, of course, a blue image. Next, uncover the other slit, and carefully sliding the second card to and fro, we can find a position (somewhere in the

yellow or orange-yellow), which *again makes the image white.*

Continuing these experiments with other colours, we find that for almost any colour near one end of the spectrum, there is another towards the other end which, with it, makes white, and is accordingly called its "complementary colour." And notice that we can do this with our *really* "pure" spectrum colours. It is not as in a former experiment (§ 48), when we divided the whole spectrum by a wedge-

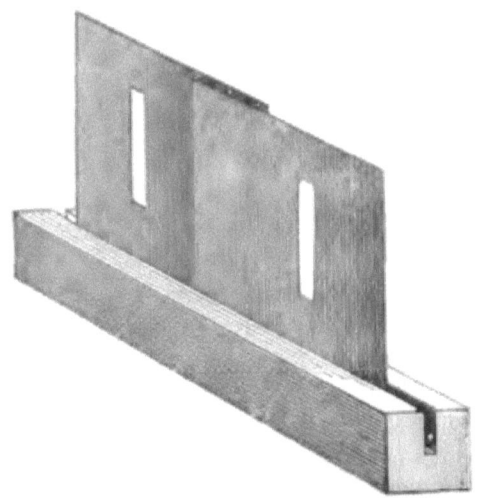

FIG. 83.—Slits for Complementary Colours.

prism into two coloured images; for those two colours really were themselves compounded, and between them contained *all* the coloured rays. But here two *single* colours make white; and hence we learn that we may have a white, undistinguishable by the eye from any other white, which will not, on prismatic analysis, yield more than two colours. We may, by continuing these experiments, find that a white may be compounded of three colours, or of more. We never get a white unless there are waves of *more than one* period; but either white, or almost any colour, may be compounded out of various constituents.

72. **The Eye unable to judge of Colour-Waves.**— Now this is another cardinal, fundamental fact. It teaches us a wonderful truth, which still more remarkable experiments will confirm; viz., that colour and light, *as we see them*, are not only purely matters of sensation, or subjective consciousness; but that this consciousness is easily deceived, and quite incapable of distinguishing between whites and colours very differently constituted. We cannot tell "by the eye" that the blue-yellow white differs from seven-colour white; nor can we tell by the eye that a compound blue, containing nearly one-half the whole spectrum, is different from a pure spectrum blue, which may be found of the same apparent shade.

73. **Mixtures of Light and of Pigments.**—And there is another strange thing. The old artists always considered that blue and yellow and red were "primary" or simple colours, and that blue and yellow made green. But here are blue and yellow, and instead of making green, they make white! How is it they ever make green? To solve the question, we fill one of our glass cells with solution of picric acid—apparently a pure yellow—and hold it in front of half our slit, as in previous absorption experiments. It allows not only yellow to pass, but *nearly as much green*, absorbing all the other rays. We take a rich blue glass, and analyse that in the same way. This apparently pure blue allows blue, and *as much green*, to pass, absorbing nearly all the rest. This itself is strange enough—that when we add green to both blue and yellow we should be unable to detect it by the eye. But it is obvious now that if we place both colours in the path of the beam, one *after* the other, the yellow solution will stop the blue rays which get through the blue

glass, and the blue glass will stop the **yellow**; but *both allow the green to pass*, and **the** net result is therefore that colour. It is the same with powders and paints; the light penetrates some little distance among their particles (§ 70), and is absorbed in the same way; and the green we get is the survival, or net result after the absorption by both, mixed with some white light reflected unchanged from the outer surface.

But get another lantern, or use both nozzles of the biunial if such is employed, and place the ordinary objectives **on** both. Place in the ordinary slide-stage of each a black card, with an aperture which shows, say, **a** 2-feet disc on the screen; arrange that the two discs **partly** overlap, and hold in front **of** each **objective one of these same two** colours, or the **blue** may **be a cell of** the sulphate **of** copper. In this way, it is obvious that instead of **the light** on the screen being the *remainder* of **two** successive **sub**tractions or absorptions, where **the discs** overlap it is the light from both colours *added*. And the result now from these two same colours is not green, but—white.[1] Analogous effects may be shown with a good grass-green glass, chosen by trial, and the beautiful purple solution of permanganate of potash.

As some may still have **difficulty** in realising that successive absorptions are the real cause of our ordinarily getting green by combining blue with yellow, or orange-yellow, we may make another striking experiment, which would seem to be a crucial test. Our previous blue and yellow allowed green to pass through each, as shown **by** spectrum analysis, and therefore green was what may be called the sole

---

[1] It is possible to get a solution of ammoniated sulphate of copper so pure a blue **as** to transmit hardly any green, when the green half of the experiment, with that particular solution, would naturally fail; but get either solutions or glasses of a good blue, which transmit green as well **as** their blue and yellow, and the same materials will do for both.

"surviving" colour. But, by search, we may find solutions which will give very similar colours to the eye, but of another prismatic character. Make a solution of oxide of copper in liquor of ammonia. This, too, is blue, and its spectrum A C (Fig. 84) shows that it allows to pass, beside blue, nearly all the green; in fact, all the spectrum to the point D. Make another solution of bichromate of potash, which is a deep or orange-yellow. This allows the red and yellow end of the spectrum to pass, from the point E, with only a trace of green, if any; and by a little dilution of one or other solution, or the use of a wedge-shaped glass cell for each, the amounts of these two colouring matters can with care

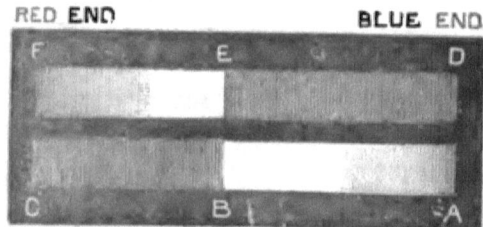

FIG. 84.—Complementary Absorptions.

be so adjusted that the spectrum of one begins about *where the other ends*, and there is no portion transmitted by both. Now here are a blue and a yellow very similar to the preceding; and their discs, when overlapped, produce, like them, a fair white, or nearly white. But superpose these two cells across the nozzle of the same lantern, and we get no longer green as we did before, but black; the two stop the light altogether.

These experiments explain a fact familiar to painters in water-colours, which as a rule are more or less transparent colours, showing very largely by the white light of the paper reflected *through* them; as may be seen by their colour on the paper differing widely from that of the cake. Hence brilliancy can only be got by a single wash; every successive wash stops out more and more of the white light; and several washes, of the proper spectral colours, instead of producing white as in the Newton's disc (§ 49), or as even a mixture of colours in *one* wash sometimes will, rather produces a muddy grey approaching black. The final result can only be the colour which is allowed to pass by all the washes, which is very little. As Sir John Herschel expresses it, the water-colour painter [1] really works chiefly by *destroying* colour, and therefore uses as few washes as he can.

Even with pigments painted in water-colour on white paper, the same facts may be proved by proper arrangements. If we mix on the palette bright cobalt-blue and the lighter chrome-yellow, a wash with the mixture gives us a good green, by the double absorptions among the particles already described. But so long ago as 1839 Mile found, that if rather narrow and long stripes were painted contiguously of each alternate colour separately, and then blended on the retina by removing the eye to a proper distance, the result was not green, but either a white or rather yellowish-white, according to the shades of yellow and blue.[2] Or the colours may be blended in larger patches

---

[1] What is here said applies far less to oil painting, which deals with more solid layers of pigment, unaided by a white background.

[2] I have seen very recently a statement by a "practical artist," who accompanied it with much hard language about "scientific theorizers," that such stripes when so blended gave *green*. All that can be said about such an assertion is, that as a general rule, with really good blues or yellows, it is simply not the fact: the statement is due to sheer ignorance and lack of experimental investigation. Some blues and yellows are so *saturated* with green in addition to the blue and yellow, that after the two latter have combined into a white, the green heavily

by a double-image prism,[1] or in other ways. Any tolerably pure blue and yellow will always, when their coloured images are *added*, produce white **or** near **it, and** cannot *anyhow* when so added be made to produce green.

74. **Primary Colour Sensations.**—Nevertheless it **is** considered probable that there really are three primary colour sensations, though different from **the** blue, yellow, and red of the old **artists.** Helmholtz and Maxwell believe the three primaries rather to be violet, green, and red. Certain mixtures of violet and green can be **made to give a** blue. which accounts for nearly the whole half of the spectrum from the blue end, **when combined,** appearing of that colour; **and red and green will also give a** yellow—most mixtures, however, giving one of an orange **shade.** Yellow appears to the eye **such a " pure "** colour, that it is difficult **to** believe it can really **be** compounded. We have seen already, however, that it will bear mixture with a very large quantity of green without the eye detecting that mixture; and it is easy to show by experiment that red and green will produce it. One method is a very pretty one easily demonstrated **by any** double lantern.[2] From one nozzle project **a spectrum by** any of the arrangements which have been described; from the other focus on the screen the image of a perpendicular slit in a black card in the ordinary slide-stage, long enough **for the** image to project some obvious distance **above and beneath the spectrum,** when thrown upon it. **Place in the stage with it a pretty** pure

predominates. Such will of course give green; **but** approximately pure and bright blues and yellows cannot be made to do so; and examination of the blue-greens which **do,** through a prism, will reveal **at** once **the great** preponderance of green which causes the exception.

[1] See § 112.

[2] **To the best of my belief it** was first so performed by Professor Tyndall.

red glass, and move the nozzle so that the red image may travel along the spectrum—somewhere in the green we shall get a fair *yellow*. Again, take a green glass with the slit, and somewhere in the red we shall again get a yellow.

Another method is due to Lord Rayleigh. A film of blue gelatine stained with litmus is placed between two glasses; prismatic analysis by methods already described, shows that it cuts out all the yellow and orange rays. A similar yellow film coloured with aurine cuts out all the blue and violet. Both together, it will be seen, stop out all but the red and green.[1] Now take away the prism, and let the light from the lantern pass direct through both, and we get an *orange-yellow*. So that here we actually have apparently blue and yellow glasses producing neither green, as in one previous experiment, nor white, as in another, nor black, as in another—but by successive absorptions, *orange-yellow!* And in all four cases prismatic analysis of the glasses or other coloured substances separately, perfectly accounts for all the phenomena.

But observe, that we say three primary colour *sensations*, and not three primary *colours*. The distinction is very important. So far as the actual spectrum and spectral colours go, even Newton's seven do not represent the case—every point in the spectrum differs somewhat in shade from its neighbours, and each one has its own distinct period of vibration, on which the colour (and other properties also) depends. No one is any more "pure" or "primary" than the other. But there are generally believed to be in the retina of an ordinary eye three *main sets* of nerves, or of the fine rods already referred to (§ 61), or whatever else receives the impacts of the ether-waves and translates them into consciousness; one responding mainly to violet, another to green, and another to red. But tuning-

[1] Both glasses, ready prepared, can be obtained of Mr. Browning.

forks, to take one obvious analogy, will respond in a less degree to *other* than their own proper notes; and so it is supposed these **sets of nerves, rods, or** other **mechanism also** respond in **less** degree to other wave periods than **their** own. It is then easily conceivable that periods which **give,** let us say, a pure spectral yellow, should also act on the brain by partially exciting the red and the green rods; and we should of course expect that **if these red** and green rods were simultaneously excited **by their own** proper colours, they would then produce the **same sensation of** yellow, or nearly so, as in the other case. The same reasoning would apply to other colours; and will **account for blue and yellow alone** making white. **For if the** blue **waves excite** the **violet, blue, and some of the green,** while the yellow waves **excite the green** and the red, **the two together set in motion more or less the apparatus which** responds **to all the colours of the** spectrum. **However the** exact details may be worked **out, the** remarkable phenomena **of** colour-blindness, and the fact that they are almost entirely confined to blue, green, **and red** colours, make it very probable, if not certain, that in the main this view is the true one.

75. **Colour merely a** Sensation.—The obvious and striking consequence at once results from such a theory, that colour is merely a *sensation.* We have already made many experiments which confirm this view, **and show that the** sensations are by no means **trustworthy guides; but** we can **now** demonstrate **that it** is so by still more striking experiments, **whose nature is easily understood.** We have (§ 61) **supposed vision to be** excited **by motion** communicated **transversely to the ends of fine rods, with which** the retina **is studded, and so communicated through the** nerves to the brain. **Now, if we press a rather blunt** pin-point on any **part of the body, or** excite sensation in any other way, we **feel it at first very vividly;** but **by degrees** the sensation

COLOUR A SENSATION.

deadens, and we take no notice. The nerves which respond to that particular feeling are by exercise, for the time, *fatigued*—tired out—and can no longer do their work. That is the reason we wear our clothes without feeling them, and of many similar facts. Now if we suppose some of the rods to respond, like tuning-forks, to certain vibrations or colours, and other rods to others, we ought to expect, under similar circumstances, results of the same kind. This is our next step in the demonstration, and the object of our next experiments. Two lanterns may be used, one

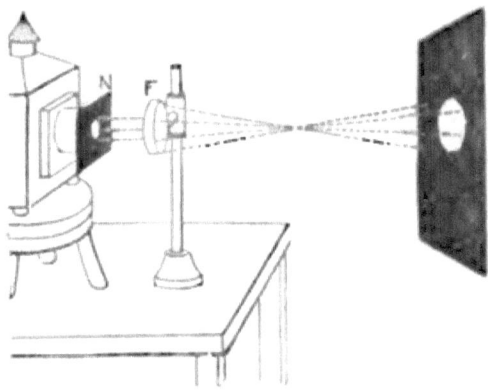

FIG. 85.—Subjective Colours.

to illuminate the screen the moment the other is shut off; but one is to be preferred, as more certain, for the first two experiments at least. Remove the objective, and prepare a black card 3 inches or 4 inches square, with a circular hole in the centre, which, when held against the flange-nozzle, as at N, Fig. 85, and there focused on the screen, gives a disc of about 18 inches diameter, or 12 inches for a short screen distance. Arrange the loose focusing lens F in front, to focus it accordingly, as in Fig. 85, and of course when the card is removed the *whole* screen is instantly illuminated. Have ready also a piece of good **red** glass, the size of the card, the picric acid cell, and a blue glass, or the ammoniated copper cell. First we **hold the plain** card over the nozzle, as shown in the figure, while we count twenty rather deliberately, fixing the eyes meantime intently on the *same point* in the bright disc. After twenty or twenty-five seconds remove the card suddenly, and where the disc was we now **see** a *dark* circle on the illuminated screen. The exhausted nerves no longer respond to the stimulus of the white screen, as do those over the untired area; **and hence, though** all the screen **is equally white, where** the disc was it appears dark.

Repeat the experiment **with the red glass held** over the card, removing both together. Here **the fibres or** nerves which respond to red vibrations alone **are fatigued; the others are not.** Hence when the glass and **card are** withdrawn, and **the screen** illuminated with white **light, the** red rays of that **light can no longer** excite in the **tired nerves** such **vivid sensations as the** other colours which **affect fresh and** rested nerves; **and so, after a second or** two, **the place where** the disc was appears *green*, though **no green light is really there,** except **as a** component of **the white.** In the same way, hold the picric acid cell in **front of the card for twenty seconds,** and we get a spectral blue, while the blue solution gives **us a yellow.** We "see" a colour which ***does not exist,*** except **in** our nervous sensations.

There is one **still more striking experiment** of the same class—that of projecting upon the screen the *entire spectrum.*[1]

---

[1] I believe **this** beautiful experiment was first performed by Professor Tyndall at **Glasgow.** Seeing the expressions of misgiving with which he introduced **it even** with the electric lantern, I was somewhat surprised to **find that** with the lime-light the effect is all that can be desired; and that with a screen distance of about 6 feet, and a good bisulphide

Arrange the bisulphide prism as for so many previous experiments, but using now as wide a slit as will give fairly pure spectrum colours; and arrange that a tolerably brilliant gas-light can be turned up instantaneously at a given signal. Project the spectrum on the otherwise dark screen, **but** in this case count thirty; and be sure vision is fixed, **by** placing a small black mark to look at about the middle **of** the spectrum. At the word "thirty," cover the lantern-nozzle, while an assistant turns up the gas; and **we see on** the screen the *complementary spectrum*, solely **due** to fatigue of the organs of vision.[1]

If further proof be needed **of** the distinction between the physical realities which underlie light and colour, and our purely sensational consciousness of **them, it** is **at** hand. Some people are more or less "colour-blind," **while yet the** perfect optical images must be formed on the retina. Some few have absolutely **no sense but that of** light and shade, though the physical reality is the same for them as for us. What the world appears to them, we may demonstrate by lighting our room with a Bunsen burner, in the flame of which dry carbonate of soda is held, or still better, a morsel of sodium in a spoon. All is mere light and shade; and we can see, as we turn up the ordinary gas, what we should miss, without the colour-sense, from our beautiful world. Finally, a large dose of the medicine *santonine*

---

prism (with a glass one the spectrum is not long enough), it can be satisfactorily shown even with an Argand gas-burner, on one condition —that the screen is not too brilliantly illuminated afterwards. Hence the illumination of the screen by an ordinary gas-light during the second stage, instead of throwing upon it the full beam from the lantern.

[1] There are a few with whom these experiments do not succeed. Some of these are colour-blind, while others seem persistently "unable" in all such matters; and the whole nervous system of some is so vigorous, that the retina does not readily become fatigued; but nine-tenths of any average audience find no difficulty.

affects the colour-sense considerably, and, besides distorting other colours, makes nearly all persons incapable of perceiving violet and purple. This strange fact is easily accounted for if we conceive that the drug renders the rods or fibres attuned to the quicker vibrations so relaxed, that for the time they only respond to slower ones. There are even some individuals (the writer is one) to whom the very same coloured object appears of perceptibly different colours or shades of colour as regards the two eyes.[1]

[1] In my case a purple object appears perceptibly bluer to one eye and redder to the other.

CHAPTER VII.

SPECTRUM ANALYSIS.

Absorption Spectra—Their use in Analysis—Continuous Spectra—The Solar Spectrum—Line Spectra—Reversed Lines—Radiation and Absorption Reciprocal—Fraunhofer's Lines—Reversed Solar Lines—Thickened Lines—Solar, Stellar, and Planetary Chemistry.

76. **Absorption Spectra.**—Some of the experiments in the preceding chapter have really been experiments in spectrum analysis, of precisely the same nature as are half of the experiments made by microscopists and chemists every day. A large part of their work consists in the examination of mere "absorption spectra" such as we have already seen upon our screen. We may demonstrate its nature by many homely and instructive experiments. Filling, for instance, our glass cells with known samples of genuine claret or other wine, we obtain by the method already described, a spectrum with certain dark bands. Now we can easily obtain other solutions, compounded with more or less alcohol, and coloured with various substances, which, "to the eye," are of exactly the same colour. But the imitative solution will not give the same absorption spectrum. It *cannot be made* to do so, inasmuch as the vibrations absorbed depend on the synchronal vibrations of the very molecules themselves. It is needless to give more examples, or to explain how we have even in this method of analysis a powerful

and delicate means of detecting adulteration, in any substance which admits of being presented as a coloured solution, or any other transparent form. We have only to ascertain the absorption spectrum of a sample of known purity; and the spectrum of the sample to be tested will at once reveal if it be genuine or adulterated.

Again, if healthy blood be somewhat diluted with water, a characteristic absorption spectrum will be observed with all the blue end cut out, and two broad bands in the yellow

FIG. 86.—(1) Iodine Vapour. (2) Nitrous Gas.

and green, near the D and E lines to be presently explained. But if we now hold in front of the slit a trough filled with blood *poisoned* by almost anything—say by carbonic oxide, or prussic acid—at once we shall perceive a marked difference in the spectrum. Light will be a Revealer of the poisoning to which the blood has been subjected.

Any coloured gases or vapours contained in closed tubes will give similar and very characteristic absorption spectra. A tube containing some iodine, vapourised over a lamp and held in front of the slit, gives a very beautiful spectrum; as also does one filled with nitrous gas (Fig. 86). The latter gas can be easily produced by putting some copper-turnings in a test-tube and pouring on them some nitric acid. Carefully avoid inhaling the fumes.

77. **Continuous Spectra.**—But, however much we disperse the spectrum of our lime-light or gas-burner, from the narrowest slit, **we** fail to find any dark bands in that; it is an unbroken band **of** colours, insensibly shading into one another (Plate **II. A**). Such is called a continuous spectrum, and it is found that *any* body which can be heated to incandescence without being vapourised—that is, which glows while retaining a solid or liquid form—gives this kind of spectrum. If we heat a piece of iron, for instance, it gives out first only *red* light—the longest and slowest waves. As we heat it more, yellow is added; then gradually green and blue; but the spectrum is always *continuous* ***as far*** *as it goes.* **A** gas flame does not give us by ordinary methods the spectrum **of** a vapour, but that of the *solid* particles of carbon in a state of incandescence: hence the continuous spectrum. This rule has been found universal. Many substances are volatilised before they can be made to give a complete spectrum: but if they can be heated so as to emit light at all, it is an unbroken spectrum so far as it extends, and it commences from the red end. Thus the body emits the very *slowest* vibrations first, gradually acquiring the quicker ones; which we recognise in popular language when we say it first becomes red-hot and then white-hot. **A** thermometer moved in the spectrum will show us that still longer and slower waves than the visible red waves extend beyond the red end, and are even hotter than the red. So that a body, heated, first acquires comparatively slow vibrations, which are **too** slow and long in their waves to excite vision; gradually **it** adds quicker and quicker ones; till at a certain point, different for each body, the motion of the molecules is so rapid as to overcome the attractive forces, and the molecules fly apart in vapour.

78. The Solar Spectrum.—We cannot show the solar spectrum in the lantern, but every student should see the leading phenomena for himself. Provide sufficient black cards to go all across a window as a black band about a foot high, resting on the middle sash or bar; they need not be fastened, as the only object is, for the width of that band, to stop out the light, and a little overlap of loose sheets will effect this. In one of the middle sheets cut a slit not more than 1 mm. ($\frac{1}{25}$ of an inch) wide and two or three inches long; and choosing a day when the sky is brilliantly lighted, and the slit standing out against it, or at least against a bright white cloud, take the bi-sulphide prism, and observe the slit through it from the other side of the room, or from a distance of not less than eight or ten feet.[1] The spectrum will be plainly seen to be crossed by several well-marked dark lines, as represented in Plate II., Fig. D. By employing more prisms to increase the dispersion, and examining the image with a telescope, these lines are increased to hundreds; but those in the plate can be seen with the single prism-bottle and nothing else.[2]

Now we have the best of reasons to believe that the sun is incandescent: his amazing heat alone makes the supposition a necessity; and his spectrum is moreover, to the eye alone, unless widely dispersed, so very *near* to a continuous spectrum as to make it almost certain upon that ground also, that the light emitted from him originally must be from an incandescent fluid or solid. What, then, is the cause of these dark lines? The probability would appear to be,

---

[1] It does not seem generally known that the principal lines of the solar spectrum can be perfectly well seen by this simple means. Even with a "lustre" of pretty good flint glass, I have never failed to see the D line, and the chief line in the blue.

[2] For further details of spectroscopes and spectroscope work, see *The Spectroscope*, by J. N. Lockyer, F.R.S. (*Nature* Series). This chapter only deals with the physical outlines of the subject.

even from what we have seen already, that they are in some way due to absorption.

79. **Line Spectra.**—We examine next the spectra of *coloured* flames, or flames which contain the *vapour* of solid bodies. The metal sodium is a convenient substance, being so easily volatilised, and the vapour so readily made incandescent by a moderate temperature. Slide back the burner or lime jet, and introduce at about the point it usually occupies a short Bunsen burner. In the flame of this hold in a small platinum spoon a pellet of sodium. The spectrum arrangements being all ready as before, we see on the screen simply a *bright yellow line*—the characteristic line of sodium, known as the D line or lines. All the rest is dark; the sodium-vapour gives a pure yellow light. (Plate II., Fig. F.)

Now, it is found that all incandescent gases give such "line" spectra; as if, when their molecules of matter were so dissociated as to be able to behave independently, they had their own periods of vibration, like pendulums of a fixed length. Some give more lines than others—sodium itself gives an additional line with the more intense heat of the electric arc; while with wide dispersion the yellow line itself splits into two lines close together. But incandescent gases give *line spectra*, and no gas or vapour gives the same lines as any other; so that when Mr. Crookes some years ago found in the spectrum of some lead-refuse volatilised, a new green line none of the known metals had yielded, he knew he had something before him hitherto unknown, and pursued his investigation till he had separated the new metal Thallium. As in all previous instances, Light was to him a true Revealer of the unknown.

Lithium and Strontium are pretty easily volatilised in the shape of their chlorides, and a small quantity of these salts will generally show visible line spectra in a *good* Bunsen burner; if not, they may be carefully held in the oxygen jet. But sodium is at once the most striking and most easily manipulated example for screen work.[1]

80. **Reversed Lines.**—In 1859 Kirchoff cleared up the mystery of the solar spectrum, by ascertaining that when the vapour of sodium was *interposed* between the slit and the solar spectrum, *the* D *line was darkened*. We have seen the bright D line of sodium, as projected from the lantern.

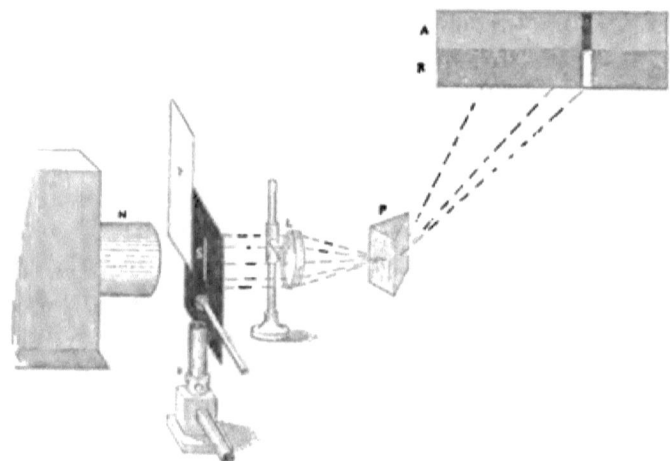

FIG. 87.—Reversed Sodium Line.

We now restore the lime cylinder to its place, and throw the continuous spectrum on the screen in the ordinary way.[2]

---

[1] With the table spectroscope, where much less illuminating power is required, many substances may be used which are useless with the lantern. With the latter, nothing beyond absorption spectra and the line spectrum of sodium can be shown without the lime-light.

[2] Nothing less than the lime-light is brilliant enough to show the reversed line on the screen. In private experiments with a prism, a good paraffin lamp is bright enough to show reversed lines with many chlorides.

As near the slit as possible we adjust the Bunsen burner, between the slit and the lens; and between the burner and the lens, again, we interpose a screen of **black** tin **or** black card, with a somewhat larger slit, so as to cut off the diffused light from the sodium from the screen. Introducing a pellet of sodium into the flame, it bursts into vivid combustion, and the light from the slit has to pass through the yellow flame. At once a *dark band in the yellow* appears on the screen (E, Plate II). It is not really black; for we know that if the sodium flame alone were employed it would give the yellow band; but it is *comparatively* dark : it stops the **larger** portion of the brilliant light from the lime cylinder.

By another and a better arrangement we can show the two spectra together, and demonstrate that in this way **we** get an exact and absolute reversal of **the** sodium **line.** It is shown in Fig. 87. Everything **but** the condensers is removed from the lantern, and a few inches from the flange-nozzle, N, is adjusted a rather large black tin screen, s, **in** which **the** slit **is cut, and** which has side-guards (not shown) in order to stop as much as practicable scattered light. As close as possible to the **slit is** arranged the Bunsen burner, B, between **the** slit **and** the lantern; by this arrangement all the light has to pass through the sodium flame, while none but what **passes** the slit can reach the screen. **On** now holding the spoon with the pellet of sodium in the flame the dark band appears **in the** spectrum; and by holding another plate, T, between the Bunsen burner and the condensers, the lime-light may be cut off from the upper portion **of** the slit, leaving the light of the sodium flame alone. The result will be as shown. One half of the spectrum will show the bright line on dark ground, giving **the** radiation spectrum of sodium, R. The other half will show the dark line on the continuous spectrum, giving the absorption spectrum, A.

81. Radiation and Absorption Reciprocal.—After all, this is what we should have expected. It simply shows us that, as we have found reason to suppose before, the molecules of matter really do take up or absorb those ether-vibrations which *synchronise* with their own vibration-periods. We form the conclusion, subject to experimental verification, that all vapours ought to *absorb* the very same colours which they *radiate* or emit when heated to incandescence. And experiment does verify the conclusion. In every case, where the vapour of a metal gives out bright lines, there, when interposed in the path of a brilliant continuous spectrum, that spectrum is crossed by dark lines.

82. Fraunhofer's Lines.—We can now perfectly understand the solar spectrum (Plate II. D). The dark sodium or D lines show us that between the incandescent body of the sun and ourselves is the vapour of sodium; other lines demonstrate the presence of incandescent hydrogen; other lines, again, those of iron. With greater dispersion the dark lines are, as already observed, multiplied to hundreds, and nearly a hundred of these are iron lines.

Fraunhofer's lines are of great value in another way. They serve as *landmarks in the spectrum*. It is difficult or impossible to determine light of a given wave-length by the colour alone; but these lines have fixed places, and, being all carefully mapped, answer every purpose. Where no solar spectrum can be employed, still the ascertained lines of iron, or sodium, &c., answer the purpose, the chief lines being all lettered and numbered.

83. Reversed Solar Lines.—All the preceding suppositions can be actually verified in the case of the sun, and we are able to obtain with proper instruments just such reversed or complementary spectra as our lantern gave us of the sodium line. If the sun is surrounded by various incandescent vapours, and these could be isolated from his

SOLAR REVERSED LINES.

overpowering **continuous** spectrum, **we** must have the *bright* lines. This was **first** accomplished during **total eclipses, when** the incandescent atmosphere gave **the bright lines of** sodium, hydrogen, iron, and other substances, **conspicuously** enough. **By** wider dispersion, which it will **easily be understood** weakens proportionately the continuous spectrum, while **not able so to** disperse the more definite vibrations of **line spectra, spectroscopists are able to show in juxtaposition the spectrum of the sun himself, and of his outer envelope of** luminous **gas, or chromosphere.** Fig. 88 shows a very

FIG. 88.—Solar F Line Reversed.

small portion of the result,[1] and we see plainly the *bright* F line in the chromosphere, while below is the portion of the continuous spectrum, **which** shows the *dark* line exactly coincident, just like the two sodium spectra in Fig. 87.

84. Thickened Lines.—There is one more rather important point. We **have** seen that solids or liquids, whose molecules **are** comparatively close, when those molecules are

[1] Taken from *The Spectroscope* (*Nature* Series).

forced into violent vibration by heat, appear so hampered as to vibrate in all periods, thus giving the continuous spectrum. When the molecules are at last driven apart, and are comparatively free, they vibrate in their own individual periods, and give lines—at least this is our hypothesis regarding the phenomena. If it be well founded, we can test it; for obviously more pressure, or compressing even gas particles closer together, ought to produce more or less *approach* towards a continuous spectrum. Experiment does verify this, and many gases have been so compressed as to give a very considerable spectrum. The easiest demonstration, however, is with our ever-useful sodium. Enclosing some fragments of sodium in an exhausted tube afterwards filled with hydrogen, and again exhausted before sealing (this is to prevent oxidation), we have the materials for a very elegant experiment.[1] We throw the lime spectrum on the screen as before, and hold the tube over the slit—there is no appearance of absorption. Applying heat, the *thin* dark line comes on the spectrum which we know so well. Continuing to apply heat, we of course increase the density of the sodium vapour, and its pressure; and as we do so the line *thickens*, till it occupies a rather conspicuous width. Removing the lamp, the phenomena are all reversed. Our theory, and the expectations formed from it, are verified to the minutest particular.

Again, referring to Fig. 88, it will be seen that the bright F line of the sun's chromosphere is much *thicker* at the bottom than the top. We gather from this optical evidence, what we know must be the case on other grounds, that the pressure of the incandescent atmosphere is much greater near the sun's surface.

85. **Solar, Stellar, and Planetary Chemistry.**—Thus we see how the spectrum enables us to ascertain with

---

[1] Due to Dr. Frankland.

wonderful accuracy much about **the** actual components, and even actual physical condition of the most distant heavenly bodies. It is interesting to find **that,** as far as **we** can trace them in our telescopes, these are constituted of precisely similar matter to what we are familiar with, governed **by** precisely the same laws. Diverse and inconceivably **far** apart—far enough for even Light, with its enormous velocity, to occupy hundreds of years in traversing the distance—all are yet one vast unity. We can trace their materials, and sort them out into groups according to their stages **of** development; we can tell if they are solid, or gaseous, and whether they have a surrounding atmosphere or not. The Light they send us is a true Revealer of all, and brings evidence of all these things in its beams.

CHAPTER VIII.

PHOSPHORESCENCE.—FLUORESCENCE.—CALORESCENCE.

Effects of Absorbed Vibrations—The Invisible Rays of the Spectrum —Three Independent Spectra non-existent — Phosphorescence — Fluorescence — Calorescence — Relation of Fluorescence to Phosphorescence.

86. Effects of Absorbed Vibrations.—In previous chapters, we have been led to adopt as our hypothesis of absorption, and of the cause of colour in coloured substances, that molecules of matter having certain periods of vibration, took up from ether-vibrations of all periods, such vibrations as synchronized with them. We are bound to ask, what becomes of these absorbed vibrations? Energy cannot be annihilated; and the motion apparently destroyed must produce certain effects. If molecules of matter thus take up vibrations from the ether, so quenching or weakening them, it may be urged that these molecules ought, in their turn, being set vibrating, to give out new light, or at least vibrations, of their own; and further it is conceivable, not to say probable, that such matter-vibrations should be excited in some measure, though not so strongly, by ether-vibrations not synchronous. Sound is our closest analogy; and we know that non-synchronal waves will set vibrating various sounding bodies. That this is so as regards heat and light, is beautifully shown by the allied phenomena of **Fluorescence** and **Phosphorescence.**

It readily appears, on reflection, that when small masses act by their motions upon large masses, the more common effect must be the conversion of quicker motions **into** slower ones. To use and expand a dynamical analogy which has been employed by Professor Stokes, **let** us consider short and choppy waves acting upon a large vessel anchored at sea. The quicker motions impart a *slower* pitching and rolling of the vessel; and these, again, cause new and ***slower waves*** in the water. But these are less perceptible than the primary waves, and may even be unnoticed, unless the water should become suddenly calm; when they would at **once** be conspicuous as long as the rolling continued, a period which would depend on the stability of the vessel. In the same way, long slow waves may more rarely **be** converted into quicker motion, and thence into *quicker* secondary waves. **Regard** the waves as motions of ether atoms, and the vessel as a molecule of matter, and **the** analogy is fairly complete.

87. **The Invisible Spectrum.**—But before we can fairly investigate these matters, we must take into our view more than the spectrum we "see" upon the screen. That spectrum has no sharply-cut ends; and we know well enough that it has other effects than visual ones. We can readily trace *heat* in it; and experiment in even a very rough way with a good thermometer soon shows us that the heat is much the greatest at the red end. If, on the other hand, we expose a photographic plate in the spectrum, we find very energetic *chemical* effects; and as regards salts of silver and many other compounds, we find that these powers of producing chemical changes are much more energetic at the violet end.

If we push our experiments further, with more delicate instruments, we find that some of the most energetic heat rays are quite *outside* of **the** visible red end, in a dark space, representing still slower vibrations than the slowest red rays. And we also find that some of the most energetic chemical effects are produced in an invisible region *outside* of the visible violet end. Moreover there are broad absorption bands and Fraunhofer lines in these invisible regions ; and there are bodies, alike in being perfectly clear and transparent to "visible" light, which differ widely in transparency as to these invisible rays. Clear rock-salt is the only body transparent to all the heat rays: and quartz is one of the most transparent to the chemical rays, which are largely absorbed by glass ; as the heat rays are almost totally stopped or absorbed by a solution of alum in water.

88. Three Independent Spectra non-existent. —Hence diagrams have been constructed showing the comparative intensity or working power of what is called the Light spectrum, the Heat spectrum, and the Chemical or Actinic spectrum ; the energy of each in every region of the spectrum being shown by a curve, whose highest point is at the place in the spectrum where the intensity is greatest. Thus, the highest luminous intensity would be over the yellow. And it has been considered that there were in a beam—say of sun-light—rays of *three distinct kinds*, called heat rays, light rays, and chemical rays.

But this is now known to be a mistake. All the rays are subject to the same laws, being reflected, refracted, diffracted, &c., exactly as the luminous rays whose phenomena we have investigated. They differ solely in their *periods* of vibration ; and their different effects are due simply to certain periods and lengths being best adapted to produce those effects. Just as with sounds, some persons can hear much graver sounds and others more acute sounds than others can, and probably insects can hear sounds inaudible to us ; so while average *visible* light waves range only between those whose lengths are from $\frac{1}{36000}$th to $\frac{1}{62000}$th of an inch in air, some persons **can** see rays **at** one end or the other, invisible to the majority. Again, it has been said that "chemical rays" or chemical effects are almost *nil* in the yellow of the spectrum; **and it** is so as regards the **salts of** silver. But the action of light upon plants is also a distinctly chemical **effect; and** this is perhaps, if anything, the most powerful **in that same** yellow region. Science knows no real distinction but periods and lengths, between any of the rays in the spectrum; each period being more or less adapted, as a rate of Motion, to produce certain effects upon the molecules of bodies, **or upon** the nerves.

Now as respect these effects, **we have a** proof that the quickest motions act most powerfully in some respects upon the molecules of matter, in the **effect of vibration** upon wrought iron. Slow motions do **not affect it**; but quicker vibrations rapidly produce a crystalline structure, showing that the molecules **are** shaken, or at least forced in some way into new positions. We see the same thing exactly in the chemical **power of** the quicker waves of light. It is almost certain **that the** atoms upon which they act **are** literally shaken into new combinations, very much as in the crystalline iron; and thus **we can see** *why* **it is** the quickest vibrations which are often most powerful in their effects. Actinism is, **in** fact, itself one of the strongest proofs of the vibratory theory of light. And as the transference of motion from the ether we here suppose, is again to the atoms of bodies, we should expect to find it in some respects most evident from the quickest waves.

We thus see, in **a** general way, what becomes of light when it meets **bodies** opaque to any given periods of vibration. It can always be traced somewhere; but is largely converted into heat **in** the body, while many vibrations which would be true visible waves as regards period, are too weak to be discerned. We should *expect* that the quicker motions would be, as a rule, most readily traced ; **and, as a rule,** converted into slower ones. And yet we ought to find some exceptions to this rule, and many proofs in one way or another of what we are supposing takes place.

89. Phosphorescence.—It will now be understood, that when we place a body in the sun for some time, and on removing it find that for a considerable time it gives out perceptible heat rays, we really have a case of what we are trying to find. Some people's eyes are sensitive to light much more faint than others can perceive: and if a large iron ball is heated white-hot, and then gradually cooled, such individuals may see the red light after others have ceased to perceive any visual phenomena. Hence there **can** be little **or no** doubt, **that in the case of** a body merely "warmed" in the sun's rays, there are also vibrations of shorter, truly visual periods, **but too feeble** for our senses. But there are quite a class of substances which, when exposed to light for a time, continue for some time after withdrawal to give out *luminous* rays, and this phenomenon is called *phosphorescence*. Prominent amongst these substances are the sulphides of calcium, strontium, and barium ; but they require to be heated and hermetically sealed in glass tubes. The diamond and fluor spar are examples in the mineral world. The compound sold as Balmain's luminous paint is one of the cheapest and best known substances. Diamond and fluor spar glow for a comparatively short time after exposure to a strong light ; but the sulphides, or Balmain's paint, will **shine for many hours by the energetic vibrations set** up in their molecules by the ether-waves.[1]

It is found **that these effects are produced** mainly, if not solely, by the quicker and shorter waves. If a sheet

---

[1] A **set of** phosphorescent tubes, which give various **colours after** exposure **to** light, **can be** obtained for **a** few shillings **of any** good optician.

of paper painted with Balmain's **paint is** made slightly luminous, and then exposed in the dark to a strong spectrum for a while, when it **is** taken into a dark room it is found the phosphorescence is *destroyed* where the slower waves fell. Those waves have the property of destroying **the** vibrations set up by the quicker waves, converting them **into** slower, or heat waves.

90. **Fluorescence.**—But there is another class of bodies which are acted upon in a somewhat similar and yet somewhat different manner. The vibrations of the infinitely small ether-atoms set up in their heavier molecules *slower* vibrations, as in the case of our ship (§ 86.) We have had one example of this when light-rays are absorbed and cause heat-rays to be emitted from the body; **but similarly, the** very quick and short invisible violet **rays may be** converted into slower *visible* **rays. Now** Professor Stokes found that when he employed quartz lenses and prisms (§ 87), the invisible spectrum at the violet end was *six times as long as the whole visible spectrum.* **It is no** wonder, therefore, that this should be the most frequent of all these allied phenomena. It is called Fluorescence, and has been specially investigated by Professor Stokes, and since by Professor Lommel.

The conversion of invisible rays into visible ones is not very well adapted for ordinary lantern arrangements, for two reasons: firstly, incandescent lime is not rich in the invisible violet waves;[1] and secondly, glass lenses, and especially the bisulphide prism, are powerful absorbents of them. The electric light is extremely rich in these rays, much more so than that of the sun. The most convenient light for ordinary lantern experiments **is** that of magnesium

---

[1] I have seen very fair actinic effects from a magnesia cylinder. Such cylinders are far too soft and rapidly consumed for ordinary lime-light work, and are therefore **not** made now; but I am inclined to think they might be so made as to give a good actinic spectrum.

ribbon, also rich in violet rays; and it may be used without any special expense, by adjusting a bit of brass tube and passing through it two or three ribbons; one of the three will then keep the others alight, but the light must be watched through a bit of smoked glass. The lantern itself and the spectrum are, however, only needed to show the creation or conversion of the invisible spectrum into visible rays; and this is best done by adjusting a spectrum with the *glass* prism on the screen. We project the spectrum, stop off all the brighter portion, and pin over the violet end and beyond it, a white card painted with several coats of a solution of sulphate of quinine in water acidulated with sulphuric acid. We see a very obvious *brightening* of the visible violet,

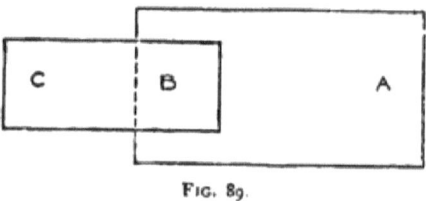

FIG. 89.

and that a perceptible region, before invisible, becomes visible where painted with the quinine (Fig. 89, B C).

Taking a cell of the quinine solution, and interposing it in the path of the lantern-beam, the screen shows that it is prefectly clear, as "clear and colourless" as water. But holding the cell so that the light *in* the cell can be seen, it glows with a beautiful bluish shimmer. Now with this cell we can show the perfect reciprocity between radiation and absorption which we have found before.

We project from the lantern, with a flint-glass (in default of quartz) train, the ordinary spectrum, A B (Fig. 89). Towards the violet end B, is attached the piece of white card painted with the acid solution of quinine, producing the brightening at B and extension to C. **Now we interpose** in front of **the** slit or lantern-nozzle the flat glass cell filled with the same solution. The B C portion is at once *absorbed*, and the spectrum brought back to its former dimensions and character. The demonstration thus is complete, that emission here also is reciprocal with absorption. We see also the reason why, in examining *cells* or *tubes* of quinine, a weaker solution **is** better. In strong solutions the effect only penetrates to a small depth, because as the light penetrates, more and more of the effective rays are absorbed; until, at last, the light that has got through a certain amount, though to all appearance as white and complete in various waves as ever, has utterly lost all power of exciting fluorescence. **The quicker** rays have been taken up by the quinine, which emits **them** on its own fresh account; and therefore they no longer exist to cause fluorescence in what quinine may be behind.

Having demonstrated the spectral relations of these fluorescing rays, however, a more convenient plan is to burn magnesium **in a small box with one side of** violet glass. The glass, of course, has no real operation beyond stopping off the more brilliant part of the spectrum, which might otherwise overpower the **more** feeble fluorescent effects. The private student may use sunlight admitted through **a** small square of blue or violet glass into a dark box. **Or if** an induction coil is at command, a vacuum-tube filled with rarefied nitrogen, or even air, gives a light feeble, **but rich** in actinic rays, which, **if** surrounded by a large tube in which fluid can be introduced, gives fine effects. The purply-blue light of burning sulphur **is** sufficient for many substances; and sulphur burning in oxygen, or some potash pyrotechnic mixtures, give powerful effects.

Substances or solutions which show good effects in the violet or ultra-violet rays, are the following:—Petroleum, which is yellow, and fluoresces blue; tincture of turmeric, or extract in castor-oil, yellow, fluoresces green; nitrate of uranium, or glass coloured with uranium, fluoresces a brilliant yellowish green. If some fragments of horse-chestnut bark are thrown on a jar of clean water containing some ammonia, beautiful streams of sky-blue fluorescent particles will descend; and two substances obtained from the bark—æsculine and fraxine, fluoresce beautifully, the first blue and the latter green. Tincture of stramonium fluoresces strongly by violet rays, and so does anthracene. Perhaps the most brilliant of all is thallene, a substance extracted by Professor Morton from petroleum residues; paper washed with this gives a splendid green spectrum beyond the visible violet rays, and designs painted with it show up brilliantly in almost invisible violet light obtained through a violet glass from magnesium wire. Fustic in an alum solution, and camwood in castor-oil, also fluoresce well.

In these substances the quicker waves produce vibrations lower in refrangibility, but not extraordinarily lower. Other than the blue and violet rays, however, can excite fluorescence; and Lommel considers that in some cases rays are produced of *higher* refrangibility. This has been disputed, but as regards the mere fact it seems beyond question. Make a solution of green leaves (nettle-leaves are very good), fresh, in benzol, ether, or alcohol, by macerating them for a few hours. The solution is a fine clear green, but it fluoresces brightly a splendid blood-red in *all* the rays except the very extreme red. If we interpose a cell in front of a slit so as to throw on the screen the "absorption spectrum" of the chlorophyll (Plate II. G), we can trace still more plainly the reciprocity of absorption and fluorescent radiation than with the quinine, since for every dark absorption band there is a proportionately brighter fluorescence in those particular rays. A pretty experiment is to half-fill a test-tube with the acidulated solution of sulphate of quinine, and fill up the other half with chlorophyll in ether or benzol. The bottom half is absolutely clear by transmitted light, and a beautiful blue by reflected light; the other **green** by transmitted light, and red by reflected light.

But though nearly all the rays up to extreme **red excite** fluorescence in chlorophyll, the resulting fluorescence is also red; and there is no proof of any rise in refrangibility. A solution in alcohol of the beautiful aniline dye called naphthalin red, however, while its beautiful orange-yellow fluorescence is most powerfully excited by the yellow rays about the D line (where also is, as usual, the darkest band in its absorption spectrum), responds less powerfully to almost all the rays except the red, beginning with the distinctly reddish orange. One of the best solutions, however, is a few drops of the alcoholic solution of eosin in water.[1] This beautiful red solution fluoresces a splendid green in nearly all the rays **of the spectrum,** showing brilliantly even by a gas-light, and therefore **with any** lantern. Here, therefore, we have a distinct case of the incident light producing quicker waves. Fluorescein, an allied substance, but which dissolves best in ammoniated water, is of a reddish yellow, and fluoresces a brilliant green; there are few more striking experiments than to throw the lantern beam on **a** jar of slightly ammoniated water, and scatter on the surface as much fluorescein as will lie on the point of a penknife. The jar must be placed so **as** to *reflect* the light to the audience, as well as to transmit it to the screen. The latter will show yellowish descending stripes, while the jar itself shows brilliant green forms like delicate sea-weeds.

[1] Lyon's, Judd's, or Hyde's *red ink* is **made** with eosin, and a few drops in water show this beautiful phenomenon well, Lyon's being the best. The fluorescent thermometers lately introduced are made with an alcoholic solution of eosin.

A simple experiment described by Professor Stokes, though only adapted for private performance, is very suggestive. Darken a chamber or box, except a small window of dark blue glass, such as transmits through a window, when analysed by a prism, only the violet, blue, extreme red, and perhaps a little green, but the less green the better. In the fullest light of this blue window lay a white plate or tile ; of course, on laying over this a slit cut in blackened metal or card, we see the same spectrum, only fainter. But now again we lay on the white a bit of bright scarlet flannel or cloth, so that through half the slit we see the white plate, and through the other half the scarlet, and can thus compare the spectra on again looking through the prism. We naturally expect the blue and violet part of the cloth spectrum to be nearly black, as it is, in the violet light ; but what the student probably does *not* expect, but what we shall find with many samples of scarlet, is to see the spectrum of the cloth *lengthened towards the red end*, and altogether more brilliant in the red portions than that of the white plate.

91. Calorescence.—Still further, as Professor Tyndall has shown, we can stop off all but the slowest waves—all but the *invisible* heat-rays—by passing the electric beam through a cell filled with iodine dissolved in bisulphide of carbon. If now a large quantity of these invisible or slowest waves be condensed upon platinum foil, or other suitable substances, they will produce either a red, or even a white heat, thus causing the substance to give out also the *quicker* waves of the spectrum. Here again is a rise, or exaltation of refrangibility, or the conversion of slow vibrations into quicker ones. This phenomenon Professor Tyndall has called *Calorescence.* It is the reverse of what occurs when absorbed light produces heat (§ 88).

Thus we have found all our expectations exactly fulfilled.

As a rule, the quickest **waves** are most noticeable in their effects, and, as a rule, the **extra-violet and** violet waves are converted into **slower** waves, the extra-violet becoming violet, or blue, or green, and the violet a brighter blue, and so on. **And yet we** have a few examples of the reverse in chlorophyll and naphthalin red and a few other substances. And yet, again, we have in phosphorescent substances examples of matter-molecules set in vibration by the ether-atoms so vigorously that they give out light for a considerable time. The probability is that *all* or nearly all substances fluoresce, but that in most cases the vibrations are too feeble to excite in our eyes visual effects.

92. **Relation of Fluorescence to Phosphorescence.**—It needs one more step to make the analogy complete. Phosphorescence, as **shown** by Balmain's paint, **ought to be** clearly connected with fluorescence. This step was always felt to be necessary **by** Becquerel, and it was he who surmounted it. **To ordinary** observation, **the effects** of fluorescence seem **to** cease immediately the exciting light is withdrawn, while phosphorescence lasts perhaps **for hours.** Becquerel, however, constructed **a** "Phosphoroscope," by which the illuminated fluorescent substances could be rapidly removed from the light, and he thus found **they** retained luminosity for a calculable time. His instrument **is** rather complicated; but Professor Tyndall, whose fertility **as an** experimenter is well known, in a recent lecture at the Royal Institution, used an apparatus much simpler and more elegant. A square iron lantern, A (shown in plan in Fig. 90), had on one side a slit B, through which alone the light could pass. Professor Tyndall, **of** course, used the electric arc; but magnesium will do nearly **as** well—anyway, we represent the light at C. Outside the slit, which is perpendicular and nearly the depth of the lantern, **is** mounted on a perpendicular axis, the cylinder, D, driven by a grooved pulley and cords, E E, from

a double system of multiplying wheels, so as to give swift rotation. This cylinder is painted with uranium or canary-glass, powdered, and the powder mixed as paint with some transparent vehicle. Turning the slit and cylinder towards the observer, it will be obvious that if there were no duration of luminosity, or true "phosphorescence," in the case of the fluorescent cylinder, it must appear dark; but on imparting rotation, it shines brilliantly with the characteristic green light. Thus, **then, fluorescence is** linked on to

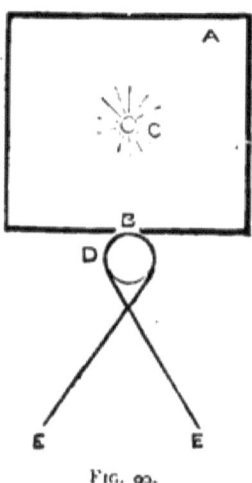

Fig. 90.

phosphorescence; and though all fluorescent substances will **not show** this with our present experimental means, there can be little doubt that it **is only a question of** degree and of powers of observation.

Once again, therefore, **Light has** revealed to us the minute, invisible **motions which** its own ether-vibrations communicate to **the molecules of bodies. Where** we may have thought all was still, it shows us molecules in constant **and** rapid **motion.** Where **we seem to** have lost that motion, further reflection and experiment yield us but another and impressive proof of the great law of the Conservation of Energy. We see that no motion is destroyed; but that every single movement does its work, and is converted into some other form. We thus get a very vivid idea of the intense *reality* of these motions, which seem hypothetical only because they elude the direct examination of our senses. That sense of their reality and definiteness will help us to understand the beautiful and new field of experiment, embracing the most splendid phenomena of physical optics, which we now have to investigate.

CHAPTER IX.

INTERFERENCE.

Net Result of Two Different Forces—Liquid and Tidal Waves—Why Single Interferences are not Traceable in Light—Interference of Sound Waves—Thin Films of Turpentine, Transparent Oxide, Soap, Water, and Air—Colour Dependent on Thickness of the Film—Proved to be Dependent also on Reflection from both Surfaces—Spectrum Analysis of Films—Soap Films and Sound Vibrations—Fresnel's Mirrors—Fresnel's Prism—Irregular Refraction—Diffraction—Gratings—Telescopic Effects—Other Simple Experiments in Diffraction—Striated Surfaces—Barton's Buttons—Nature of Interference Colours—Measurement of Waves—The Size of Molecules of Matter.

93. **Net Result of Two Different Forces or Motions.**—We have now to study a class of experiments which most of all clearly demonstrate the true wave character of the phenomena which constitute Light. We know that different separate motions can act upon one another, so as either to combine and strengthen, or to neutralize and destroy each other, simply because the actual motion of any particle must result from the net sum, difference, or other result of the forces which act upon it. Take a billiard ball travelling in a direction and at a rate resulting from some stroke of the cue; if we impart another impulse in the same direction the velocity will be increased; while if the ball be met by a second force of the same amount it is brought to a standstill.

94. Interference of Liquid Waves.

The same must result in the case of any series of vibrations of equal amplitudes and periods, such as constitute a wave. If we drop two stones at some distance apart into the same pond, the circular

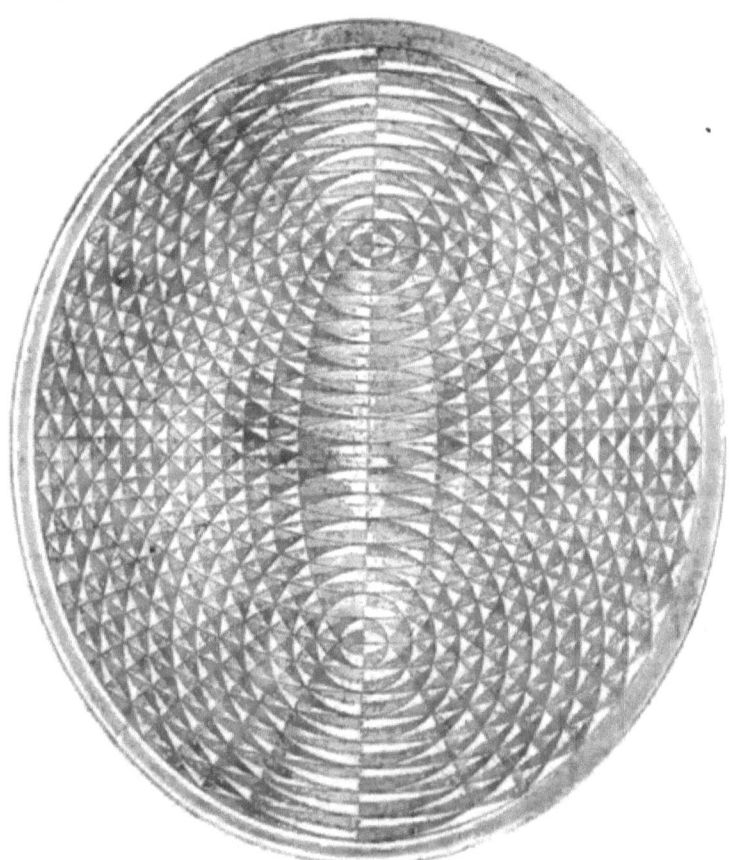

FIG. 91 —Interference of Liquid Waves.

waves from one will cross those from the other. At some points the crests will coincide, and reinforce each other; at others the same particle of water is elevated by one wave and depressed by the other; there it is at rest. The consequence is a beautiful pattern caused by the intersecting ripples. Fig. 91 shows such a pattern caused in an elliptical bath of mercury by a drop or point introduced at one of the foci. They can be beautifully shown by the vertical attachment (§ 10) to the lantern, laying over the condenser a glass plate to which is cemented a circular tin wall, making a circular tank some 6 inches in diameter and an inch deep, with a glass bottom. On focusing the surface, and then touching it with a pointed wire some distance from the centre, the intersections of the original and reflected waves will be beautifully depicted upon the screen.

95. **Interference of Tidal Waves.**—The same thing is true of tidal waves, a remarkable example of which is found in the channel between England and Ireland. The flood-tide, sweeping round from the Atlantic to the north and south of Ireland, meets about a line which usually passes just across the south of the Isle of Man. There the two *currents* destroy one another, and there is practically none, while the rise and fall of the tide is greatest. But going back from this point to north and south, there are also two points, near Portrush, in Antrim, at the north of the Irish Channel, and near Courtown, in Wexford, at the south, where the falling tide meets the next rising tide; at these points, therefore, there is practically no rise or fall of tide whatever, while the *current* is at the maximum. The same is true of the vast tidal waves that sweep round the globe. At certain times the sun-wave agrees with the moon-wave, and then we have the greatest tidal motion; at others the sun's wave opposes the moon's wave, and we have the least motion.

96. **Single Interferences not Traceable in Light and Sound.**—But here we must make a very important distinction, the want of which has caused many a student difficulty. In the foregoing cases we could trace the interferences

of *single waves*, because their motions were large to us, occupied considerable time, and thus enabled us to trace them—most clearly in the grand tidal waves, which **are longest of all**. The student is at first apt to fancy **that, in a similar** way, rays from **any two** points of light must be constantly destroying one another by interference, much as in **Fig. 92**, supposed to represent the rays from two lighthouses. **And** to some extent they undoubtedly do so. But they can only thus **act** on each other at **the** detached *points* where the undulations cross; and in the case of light and sound the vibrations are so enormously rapid and numerous that

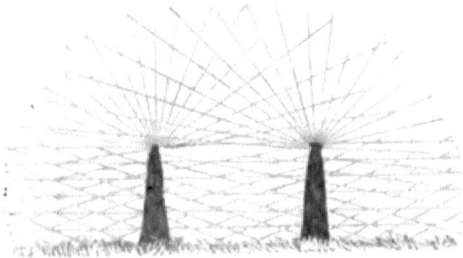

FIG. 92.—Two Lighthouses.

comparatively few extinctions of this kind are not sensibly missed.

But if we can bring a whole *wave series* to act upon another exactly similar whole wave series, then any effect at one point in any wave of the series is repeated throughout the series, and the effect becomes visible. In the case of sound we **can** get similar wave series pretty easily by employing exact unisons; and so it will be found, if a tuning-fork be struck and held close to the ear, that on turning it round on the stem there is a position in which the sound is nearly or quite extinguished. This position differs, as it should do, with the key of the fork, but with either an A or **a** C fork is when the two prongs are at an angle of nearly 45° with the direction of the ear. If the fork is steadily rotated, the sound will be alternately extinguished and reinforced, according to the phases in which the waves from each prong encounter one another.

A moment's reflection shows us that in the case of light, as a rule, we can only get the exact similarity necessary by employing two beams of light from the same original point of emission, or very nearly so; but if we can bring two such exactly similar series of waves again together, or so *close* together that the ether-atoms set in motion by each can act upon each other, while the paths of the rays are sufficiently parallel for many successive undulations to come into the same relations, then we ought to get effects which shall be visible to us. There are several methods of effecting this object.

97. **Colours of Thin Films.**—The simplest and one of the most striking is reflection from a "thin film." If a pencil of light A strikes any thin transparent film at B, we know that a large part is reflected at a similar angle to C. But the rest is refracted to D, where (unless at the angle of total reflection) a portion passes through and is lost to us, while another portion is reflected to E and thence refracted to F. It is evident the ray E F must be precisely similar to the ray B C in the periods of its waves, and also precisely *parallel* to it; and, if the film be thin enough, it should also be *near* enough to it to cause interference. As to the phenomena we ought to expect, remembering that every colour has its own wave-length, and reverting to the wave-slide shown in Fig. 71, we see there how the retardation of the central section of that slide by a given distance brings the long waves into contrary phases, while the short waves of half the length, at the same time, exactly coincide. A very little thought shows that with waves of all various lengths, only

one length can be **exactly** coincident and only one *exactly* contrary in phase, when one **set** is retarded a given distance; all others being affected one way **or the other** in varying degrees. Applying this to colours, and remembering what we have **found already as** to the effect of suppressing **any** part **of** the spectrum, we therefore expect that *colour* **will** be produced. **Now** in Fig. 93 this is what we have. **The** ray E F **has had** to traverse the film twice, from B to D **and** from D **to** E, before it **can** start on **its** journey parallel with B C. It has got by **that** distance *behind* B C, and **as** this retardation affects each colour differently, **and more or less** suppresses some colours while it **more or** less strengthens

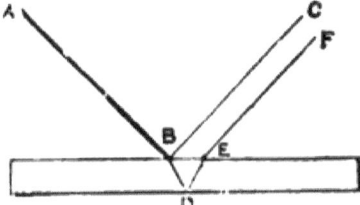

FIG. 93 — Reflections from a Film.

others, we are prepared **to** expect colour if the film is thin enough to allow the **two to act upon** each other at all. We can subject **the** matter to experiment in many ways, **all** of which give phenomena of great beauty.

Take a small black tray or hand-waiter W, say 8 by 12 inches, lay it on the table, or on a block to raise it if needful, and fill it half an inch or so deep with water. If it is a lime-light, cant up the back of the lantern, **as** in Fig. 94, so that the parallel (or rather *slightly* divergent) beam from the flange-nozzle with the objective removed may fall on the water at W, and be reflected to the screen S. If it is a gasburner, the reflector must be used to bring the beam down; and in some situations it **will** be best to *focus* the surface on the ceiling, as in some previous experiments; but with the lime-light the plain beam is best. Having adjusted all, dip the end of a pen-holder or any pointed rod into a bottle of spirits of turpentine, and let a single drop fall on the water.

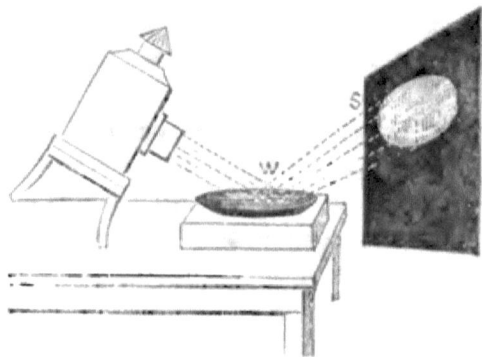

FIG. 94.—Film of Turpentine.

It spreads out instantly, and the reflected light on the ceiling or screen is tinged with the most beautiful colours.

Support a polished steel plate 3 or 4 inches square on a small tripod, and place a spirit-lamp or Bunsen flame under-

FIG. 95.—Film of Oxide.

neath the centre (Fig. 95). Bring down the light from the lantern, and focus, as in Fig. 26, then light the lamp. As the film of transparent oxide forms on the steel, colour appears, which gradually takes the rough shape of variously

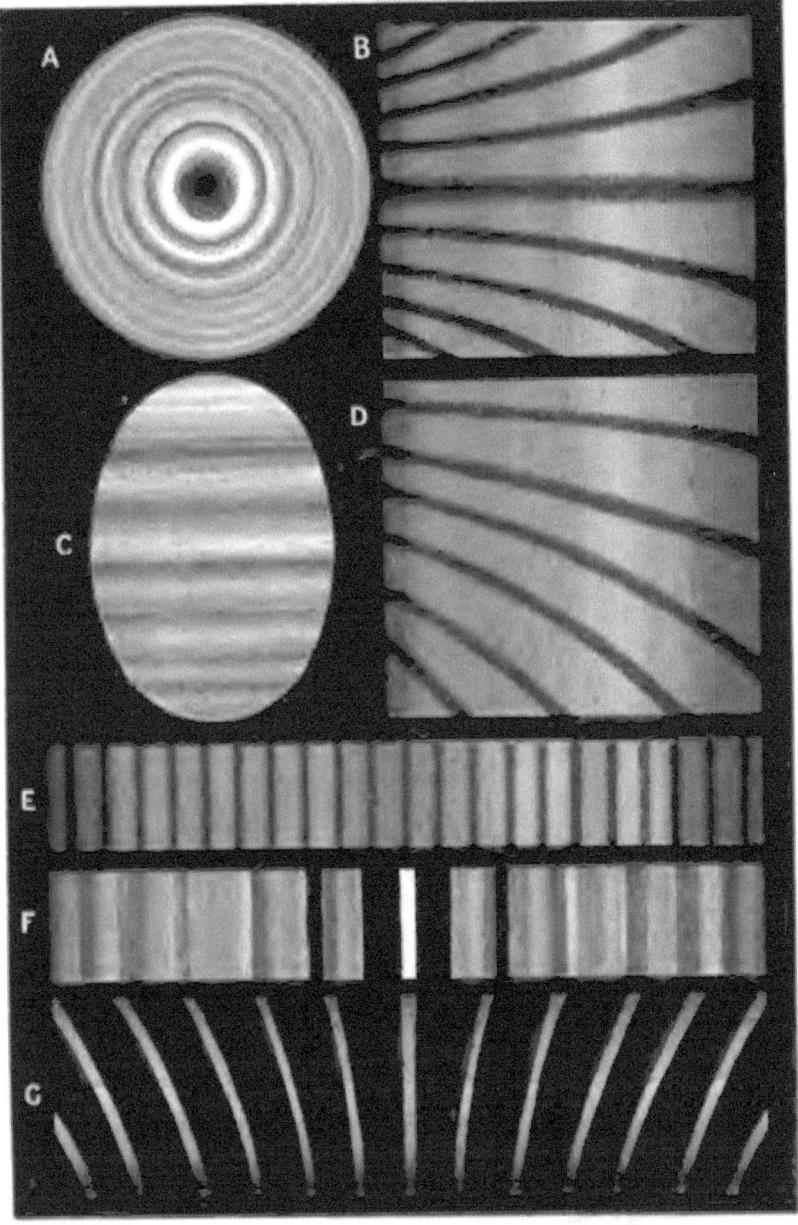

A   Newton's Rings   B   Spectrum of Ditto.
C   Soap Film   D   Spectrum of Ditto.
E   Spectrum of Light reflected from film of Mica
F   Diffraction Spectra of slit, through a Nobert's grating
G   Ditto observed through Prism.

coloured rings, though **not very regular.** This experiment is a little tedious; but it is a very interesting one. After the lantern and plate are first adjusted, some other experiment may be proceeded with while the heat operates; or oxidation may be hastened by covering the hot plate with a film of solid paraffin, which will easily spread and greatly hasten the process.

98. **Soap Films.**—Soap films, however, offer the most splendid phenomena. A good solution is of great importance, and there are many recipes, the most generally known being Plateau's. For this dissolve 1 ounce oleate of soda (6*d*. per ounce), cut into thin slices, in two pints (40 ounces) *distilled* water, rather **hot.** Mix the solution with 30 fluid ounces of *pure* glycerine, and shake violently for several minutes, several times, with some hours interval between; **then** leave for several days **before** use, as the solution "tempers" together for a **certain** time, and filter clear. This recipe does well for *warm* **weather; but it is not the** best that can be made even for that, while it often fails in cold weather; in fact no stated recipe can be given for the English climate. The method **I adopt** is to make at least the quantity named, as above, and having made it pretty warm, add to it about half an ounce of shavings of Marseilles or Castile soap, which will all dissolve in the hot solution. Warm this at convenient intervals on the hob, or otherwise, for several days, shaking it and leaving it **to** settle between. Finally, let it thoroughly cool, and then filter it at about 50° through Swedish paper into stoppered bottles, which will filter out all precipitate and make it clear. If, however, the weather turns very cold, or after considerable time, it may become again turbid **and** useless, and it is necessary either to filter out the precipitate, or (what does as well in the former case) to warm the solution before use, warming also the saucer and other apparatus. The *rationale* of all this is, that a cold

M

solution does **not keep** dissolved nearly so much soap or oleate as at a moderate temperature. **The addition of** the soap toughens the film more **than** I find possible with pure oleate. So also will **a very little** gelatine; but this latter is apt to decompose after a while. After all, however, the first solution thus **made may,** very likely, not be thoroughly satisfactory. **I then** provide a number of such paraffined rings as are **described** presently, take small quantities of the above "stock," and add to them separately (making memoranda) different quantities of soap solution, or glycerine, or water, **well** shaking **and** leaving them some hours to "temper"; then stretch a film of each on the rings, the ends of which are **stuck into** horizontal bradawl-holes in a long slip of wood. Notice is then taken, comparing several trials, which lasts **the** longest; **and when that is ascertained, the** whole solution **is made up to that standard.** By this tentative method far tougher solutions may be got **than by any** recipe that can be given for a variable climate,[1] **and** with the varying

---

[1] A few other **recipes** may, nevertheless, **interest** the reader. Professor Dewar gives: **Water** 20 ozs., soap 1½ oz., glycerine 15 ozs.; but I cannot get anything to **carry so** much soap as this, if the soap is **at** all **good. Professor Mayer, of New York, gives:** Castile soap-shavings **1 oz.,** water ½ pint (10 ozs.), glycerine ¼ **pint.** This also I cannot dissolve in the water; but taking the **effective** part of it and mixing a *saturated* solution **of** Castile or **palm-oil** soap with half its bulk of glycerine, gives **a very simple formula, which will** do fair work in all **but coldish** weather. A solution used by Professor Tyndall is made by taking a **3 per cent.** solution of oleate of *potash* (*i.e.* 1 oz. oleate to 33½ ozs. water), and mixing 3 volumes of this with 2½ volumes glycerine. I have quite satisfied myself that, taking oleates and soaps as obtainable, **no formula will give such good results as** the tentative method above; though I doubt if Plateau's can be much surpassed for a temperature of 70° with pure oleate and Price's best glycerine. For a sudden emergency a simple solution of any good glycerine soap will often answer moderately well. It is always to be remembered that the *greatest* quantity of soap which can be dissolved by no means gives the best result.

qualities of soap, glycerine, and even oleate. It is perfectly easy to place on the ring-stands shortly described, bubbles nearly a foot in diameter; and I have several times **blown in** the usual way globes *half a yard* in diameter; but **even a** twelve-inch is a magnificent object. A solution must always be brilliantly *clear* to do good work, and if turbid, should be filtered through Swedish paper before any important experiments. The saucer used must also be perfectly clean.

Make a few rings of 1-16 inch iron wire, like A (Fig. 96), 2½ inches diameter, and a few rather larger, say 3 inches; the latter stick into wooden **feet as** at B. Solder the joints,

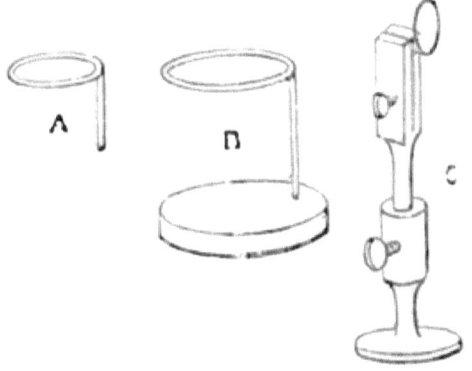

FIG. 96.—Rings for Soap Solution.

afterwards smoothing them off, and then dip the rings into melted paraffin, or warm them and smear with it, which keeps them from cutting the films. Arrange several of the B stands in a row, first dipping their paraffined rings in a saucer of solution and wetting them thoroughly with it; then we can with a little practice blow large bubbles and place on the stands. Through the whole row throw the full lantern beam, which gives a fine effect. Or a large bubble may be blown on the saucer itself, first carefully soaping it to the very rim; if this is placed in front of the lantern and the nozzle canted down towards it, fine reflections may be cast on the screen. We must avoid carefully all *froth* in the saucer, keeping as free as possible from all bubbles but the one we are blowing. A clean common tobacco-pipe may be used, but is very tedious; the brass ones sold by opticians are little better, and very nasty to use; by far the best instrument is one of the smallest *glass funnels* (an inch across), sold for filtering small quantities of fluid, on which is sprung half a yard of the smallest india-rubber tubing.

But the finest experiment is with a flat film. Pinch one of the A rings (Fig. 96) as at C, in the clip, the ring standing *above* its stem,[1] and adjust so that the plane of the ring is perpendicular, and stands the same height as the lantern nozzle. Turn off the lantern L (Fig. 97), parallel with the screen, then dip the ring in the saucer of solution, lift a film, and place it, as at A, at an angle of 45°, with the whole light concentrated on it, which will be reflected to the screen. Turn the clip-stand till the reflected light is central on the screen, and then adjust the loose focusing lens F to form an image. A glorious image it is, as band after band of interference colours travels up the oval image of the wire (the bands really move *down* the film as it becomes thinner, but the image is of course inverted), while every motion of the film from the least breath of air is pictured plainly (Plate III. C). Simple as it is, there is no more beautiful experiment than this, and the film with a good solution will last for an hour.

Any surface (not too convex) of iridescent glass (which has a film on the surface whose refractive power has been altered by a chemical process) may be focused in the same way as the soap-film.

[1] The film almost always lasts much longer than if the stem is uppermost, owing to the thinnest portion being dependent from the smooth and unbroken circular wire.

FILMS OF AIR.

Another beautiful experiment is with a film of water. Blacken a piece of glass on the back, rub a piece of soap over the surface, and clean off with a chamois leather. Pinch this in the clip, adjust it like the soap-film at 45°, and focus; but *keep it cool* by interposing one of the glass cells filled with alum solution, or the experiment will fail, as it depends on condensation of the breath by the cold surface. Then blow on the centre

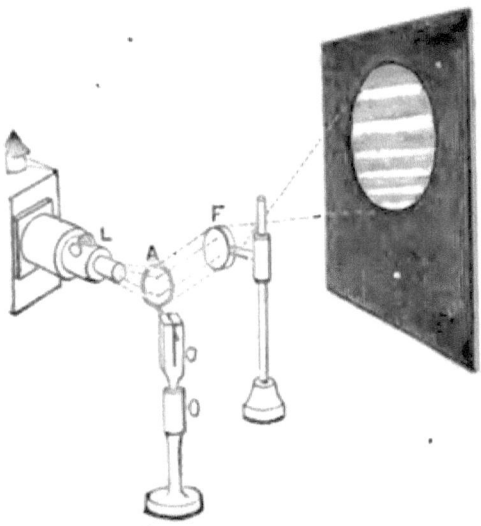

FIG. 97.—Flat Soap Film.

through an india-rubber tube of ½-inch bore. As the breath condenses roughly circular coloured rings will form, and gradually change as the moisture evaporates.

Next we may take a film of air. Buy two squares of *plate*-glass, say 3 inches square, and grind off the sharp edges to prevent scratching. Carefully clean them, and then carefully slide or grind them with moderate pressure smoothly together. We very soon see beautiful fringes of gorgeous colour. When satisfactory, pinch one lower corner of the double plate in the clip, and the three others with loose wooden spring letter-clips. Focus as before : all will be reproduced on the screen, and as we further pinch anywhere, even with the finger and thumb, changes and movements of the colours will demonstrate that the particular colour wholly depends on the thickness of the film.

FIG. 98.—Newton's Rings.

99. **Thickness of the Film.—Newton's Rings.**—We want to know, if possible, however, what that thickness is, and the last experiment probably suggested to Newton his famous "rings." He placed a convex lens of very slight convexity in contact with a flat glass, as in Fig. 98, against which it was pressed by screws. A simpler method sometimes employed is to cut two circular *flat* glasses (they must be a

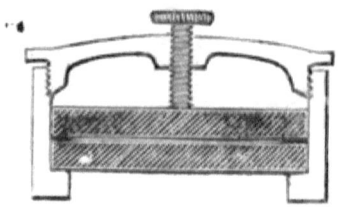

FIG. 99.—Newton's Rings with Flat Glasses.

$\frac{1}{4}$-inch thick, or, at least, one must be so), and having carefully cleaned them, place a ring of gold-leaf between them at their circumference. Mounted as in Fig. 99, pressure from the centre screw at the back produces, as in the other case, "Newton's rings," which, in either case, are presented to the lantern and focused on the screen precisely in the manner of the soap-film (Fig. 97). This method is within the power of many who like to construct their own apparatus.

It is clear that, knowing either the curve of the lens, or the thickness of the gold-leaf, it is very easy to calculate

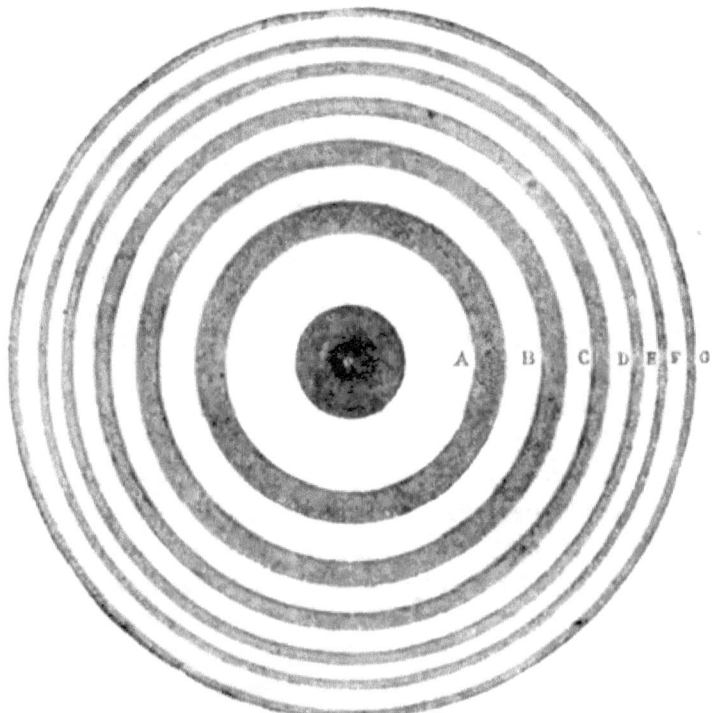

FIG. 100.—Newton's Rings.

the thickness of the film of air at any given distance from the centre. Newton found that when he employed pure monochromatic light, he obtained *recurring* rings of coloured light, or of darkness, as in A, B, C, D, E, F, Fig. 100, at once, twice, thrice, and other multiples of one definite small thickness. He soon discovered another beautiful fact, viz., that the rings were broader, or required a thicker film, in **red** light than in blue light; **and** finally, by a movable prism, **he** threw the successive **colours** of the spectrum on the rings, and found them gradually contract as he travelled towards the violet end. With most lanterns there is hardly light enough to employ **this** beautiful method; **but** the phenomena may be shown **as** follows :—Arrange the Newton's rings, and **focus on** the screen as before. Provide one of **the** movable slide-frames now used by all lantern lecturers, **and** fit in it two half-size glasses, one blue and one red. Condense the **full** light on the rings; and as close to them as possible, between them and the nozzle, hold the slide ; as **it is** moved **from** side to side, the rings will open or contract as the red or blue glass is interposed, and when they equally cover the rings, the two semi-circular segments will be seen **not** to coincide, the red being larger in diameter than the blue. It is easy to understand, therefore, **why,** if we employ white light, we must obtain rainbow-coloured circles.

100. **Test of the Emission Theory.**—Having **to** account for these phenomena, and adopting for practical purposes the Emission Theory (§ 58) as his working hypothesis,[1] Newton accounted for his bright and dark rings recurring at **every** multiple of a given thickness of the transparent film, **by** supposing that the "particles" of light suffer alternate "**fits**" **of** transmission or reflection, at *regularly recurring* **intervals or** distances. Professor Tyndall supposes that he **imagined a rotation** during their progressive motion, and this is **not improbable;** but it **is** only a supposition. If, then, light reaches the first surface of the film in a fit of transmission, it enters it and travels to the second surface ; and if the thickness is such **that** it is in the same fit or phase

---

[1] There is ample evidence in his *Optics* that Newton was very strongly attracted towards the Undulatory Theory, but did not feel justified in adopting it, owing to difficulties he was unable to **solve.**

when arrived at the second surface, it is again transmitted, and so is lost to view by reflected light. There is at that point, therefore, a dark ring; and obviously at every *multiple* of that thickness another dark ring. **If, on** the contrary, the **particle** is in the opposite or reflecting fit when it reaches the second surface, it is reflected and forms a bright ring.

It will be plain how, on either hypothesis, the "particles" or "waves" of red light are *larger* than those of blue.

We can easily, however, test the two theories. Obviously all the light that has anything to do with the rings, according to the Emission Theory, enters the first surface; and the "fit" in which it reaches the **second** has *alone* anything to do with them. **On** our wave hypothesis, it **is the** interference of waves reflected from *both* **surfaces that** causes them. **We** have not come to Polarisation yet; but it may be briefly stated that polarised light utterly refuses to be reflected from glass **at** a certain angle; and this polarised light we readily obtain by fitting a "Nicol prism" on to the nozzle of our optical objective.[1] All our light is then polarised; and when the long diameter is vertical and the Newton lenses are adjusted at an angle of 55° to 56° with the beam from the lantern, none of it will be reflected from the top glass, or in other words, from *the first surface of the film*. And when two plain glasses are used, none would be reflected from the second surface either. But metal is subject **to** quite other laws, and does reflect light copiously under such circumstances; therefore, by substituting for the bottom glass one which has **been** silvered or platinised, we can still get reflection from the second surface of the film of air. On Newton's theory, we ought therefore still to get the rings. But we do not. There is *light* on the screen, but the *rings*

---

[1] For details and explanations on these points see Chaps. X. and XI. Only sufficient is stated here for the purposes of this experiment.

have vanished, in the proper position of the Nicol; to be restored again when this is so turned round as to restore reflection from the first surface also.

Further yet; if we next adjust the lenses so as to meet the light at a still greater angle (from the normal) than that of polarisation, and thus *partially* restore reflection from the first surface, on rotating the Nicol we get a complicated and beautiful phenomenon, first discovered by Arago; viz., in one position the rings are of certain colours, and when the Nicol is rotated 90° they show *complementary* colours. Detailed explanation of this is here impossible; but it can be understood how we thus prove absolutely that the rings are due to the mutual actions of the rays of light reflected from *both* surfaces of the film. We may prove this in yet another way, by substituting for the glass and metal surface, two glasses of widely different refractive powers, whose polarising angles are therefore also different (§ 120). We can then adjust the beam of light to either, and in either case, or by destroying reflection from *either* surface of the film, we destroy the rings. This last method of demonstration is, however, only suitable for private experiment.

101. Spectrum Analysis of the Rings.—We further beautifully illustrate the matter by bringing to bear our never-failing method of spectrum analysis. Cover the pair of Newton's lenses with a disc of black paper or card, having in it a slit, say, $\frac{3}{16}$ of an inch wide, and reaching all across, exactly over the centre; the slit then crosses all the bands at right angles, and the appearance, or image on the screen when focused there, is like one of the bands in Fig. 101. The whole arrangements are shown in plan in Fig. 102. The lantern must be turned considerably away from the screen, so that the reflected beam may have a small angle of incidence; or else, as the glasses are so thick, the light from the film will not be able to emerge from the narrow slit by which it enters, and there will be only an image of a white slit as reflected from the *upper* surface of the top glass, and none of the portions of rings, which is what we want. The black disc is placed on the face of the lenses, L, with the slit perpendicular,

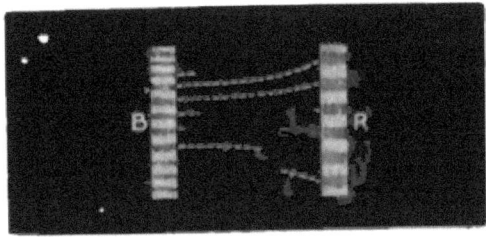

Fig. 101.

and the reflected slit is focused by the loose lens, F, at about the screen distance, but must be considerably divergent from the screen to allow for refraction by the prism, P, which gives the spectrum on the screen, S. The lenses are drawn much larger in proportion for the sake of clearness.

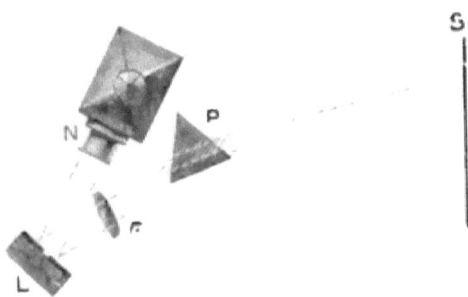

Now we know that if we use red light we get transverse bands of red and black, and that blue light gives us *narrower* bands, and closer together, of blue and black, as shown at B and R, Fig. 101. If, then, we pass the image of the slice of rings from white light through a prism, and throw its spectrum on the screen, it is easy to see what we may expect if the theory be correct. Dr. Young saw it long ago, and in his "Lectures" he has *drawn* what he foresaw clearly with the eyes of his mind, though there is no actual record that he observed it in fact, as we are about to do. Obviously we must get a spectrum of the reflected slit or slice of light. And seeing that red light gives us bands of a certain width, as at R in Fig. 101, while blue light gives us narrower bands as at B, by drawing the imaginary lines as dotted, we can see what must occur when all the colours are dispersed into their several places in the spectrum. We *must* get, unless all our theory is wrong, the beautiful appearances shown in B, Plate III.; the spectrum of the slit being crossed by parabolic dark lines which will show exactly the waves *cut out* by interference at every thickness of the film. Such a spectrum was foreseen by Dr. Young, and drawn, from theory alone; and such a spectrum with its parabolic interference bands, stands before us on our screen.[1]

The flat soap film may be analysed in precisely the same way, but is done with less trouble, there being no thick glasses to interfere with the effect. It is sufficient to arrange for a parallel beam, and place the cardboard or adjustable brass slit on the nozzle, as for previous spectrum-work. The slice of light from the nozzle will then sufficiently mark the image whose spectrum is desired, and may impinge at any convenient angle. When all is adjusted, take up a fresh film; and as it thins, the dark interference bands, showing the waves destroyed by interference, will *travel* across the spectrum steadily as long as any colour is shown.

[1] To the best of my belief this beautiful experiment was first publicly made with the lantern by Professor Tyndall.

The appearances of the film and its spectrum are shown at C, D, Plate III.

The student can observe these phenomena directly through any form of prism; and nearly all the phenomena described in this chapter can also be seen privately, without any lantern or other expensive apparatus whatever.

Towards the edges of a pair of Newton's lenses the rings of colour seen in white light disappear; and some people realise with difficulty why this is so, when films reach a certain thickness. There are two reasons. One is very much the same as the reason why we could not get a pure spectrum from a wide slit. (§ 50.) Interferences are produced, up to a certain point; but so many very narrow rings or fringes are mingled, so close upon one another, that the visual effect is white. Homogeneous light will show rings in much thicker films, and is one proof that this is so. Spectrum analysis of a film not so thin is another. With care a film of mica can be split so thin that, while it appears to reflect perfect white light, if a slit of this light be analysed, by blacking the mica all but a narrow stripe, and then treating this stripe of reflected light like the slice of light from Newton's lenses, the spectrum will be seen *crossed* by numerous straight interference bands. Such a spectrum from a film of mica is shown at E, Plate III. But when the film reaches a still greater thickness, it will also be seen on consideration that the two reflected rays into which each original ray is divided, are so far *separated* by refraction that they are no longer close enough to interfere with each other at all.

Mica films may be employed in yet another way. Project a spectrum from a slit as usual, and across the path of the rays hold the film obliquely. A portion of the rays are then transmitted direct, while a portion are reflected within the film and then transmitted, after losing a certain distance as before. Again the spectrum will be crossed by beautiful interference bands, though hardly so distinct as by the preceding method.

It is very clear that the thicknesses at which the light and dark rings occur, must give us definite information as to the *wave-lengths* of the different colours of light. This point will, however, be better explained in connection with other cases of interference (§ 110).

102. Thin Films and Sound Vibrations.—The most magnificent illustration of the colours of thin films, and the most elegant application of them in physics, must now be described. They are due to the researches of Mr. Sedley Taylor, and relate to the vibrations of telephone plates. It is well known that, by the variable attraction of a magnet under the influence of variable currents, and vibrations thus caused in a thin sheet of iron, the most complex sounds of the human voice, or other instruments, are reproduced by another sheet of iron at the other end of a telegraph wire; but it is very difficult to realise how complicated speech can be reproduced by such simple means. By stretching a soap film over an aperture in a plate laid over a resonator, and exciting the vibrations of the air contained in the latter, Mr. Taylor obtained most beautiful figures which elucidate the matter, by showing how complicated these vibrations are. Later on Mr. Tisley constructed the *phoneidoscope*, which, when sung into through an open mouthpiece on the end of a tube, shows the same phenomena in a film laid horizontally over the other end of the tube. The scale of this instrument is, however, far too small for lantern work, and it is very difficult to avoid constant bursting of the film. After many trials the following apparatus was constructed, which meets both defects, and easily gives magnificent phenomena. Its main parts are shown in Fig. 103. A is any vessel open at both ends, 2 to 3 inches across at the top, and with a neck at the bottom just large enough for **an inch** vulcanised india-rubber tube to stretch a little tightly over **it**. The article actually used was what **is** called a " gas-jar" **(cost** 10*d*.), but this is deeper than necessary, **and a piece of** a common bottle, cut round the body **and neck, and** ground flat each **end**, would **do** just as well, or a tin funnel with an elbow at **the** bottom and a flat ring soldered flush round the top would answer admirably. B **is a** brass or any other elbow, of **the same** aperture **as the** neck, to which it **is** united by a collar, C, of the india-rubber **tube** stretched **over**

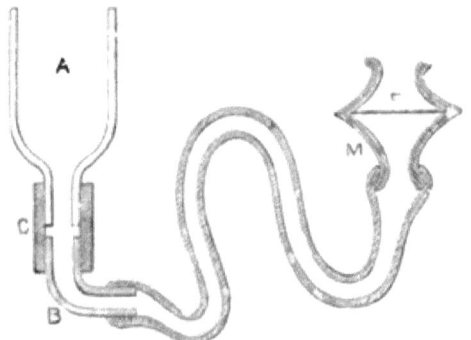

FIG. 103.—Lantern Phoneidoscope.

both. The whole is fitted **into any** thin box, the **glass** projecting through a **hole** cut in **the top.** Over the other end of the elbow is stretched any convenient length of **the** india-rubber tubing; and it is more imposing if to this is attached a sufficient length of " composition ". or other pipe, to reach the entire length of the lecture-hall, or other room. The other end of the elastic tube is stretched **over** the neck of a kind of telephone mouthpiece, M. In my first experiments, I used **a** simple open mouthpiece, like that of any speaking-tube, over which was stretched tightly a film of india-rubber (in this case a child's balloon), tied as housekeepers tie the bladder over a jam-pot. On this was inverted and held by hand a similar mouthpiece, to the narrow end of which the lips were applied. This answered sufficiently well; but I prefer a regular mouthpiece as shown, with a diaphragm of thin mica, F, $2\frac{1}{2}$ inches *clear* diameter, which can vibrate between the two funnels, and an outward opening for the lips. Either, however, will answer.

All this being put together, some discs must be prepared, of metal or very thick card, $3\frac{1}{2}$ inches or 4 inches diameter, with openings in the centre of different shapes; the best effects are from circular, square, and hexagonal apertures, which should be about $1\frac{3}{4}$ inches or 2 inches diameter. The apertures should be *bevelled* towards one side of the plate (this is done because the film will, by its contractile power, draw flat to the smallest side of the aperture), and if of metal, the discs must be carefully flattened; if of card, well varnished to resist the moisture. Finally, they are blackened, and we are ready for work. The apparatus is first adjusted, so that one of the discs laid across the top of the vessel A lies nearly level, and then lantern and all are arranged so that the aperture is truly focused on the ceiling, or other nearly horizontal screen overhead (a slight incline both focuses and shows best), precisely as the glass of water is focused in Fig. 26. Then we dip the end of a strip of card, rather wider than the apertures, in a saucer of soap solution, and drawing it carefully over the *smallest* side of an aperture, readily cover it with a film. (Use no more solution than necessary.) Several discs may be covered at once to avoid delays. Hold the disc at an incline, or upright, till interference colours begin to show; then lay it centrally on the vessel, A, with the soap *downwards*, so that a *dead black* margin surrounds the film. Of course, if all is right, we have an image of the film on the top screen. If not, be sure all *is* focused properly, and that all light possible is condensed upon the film. Then take the mouthpiece, or let some one else do so at the other end of the room, and sing into it. Not only every note, but *every different vocal sound on the same note* will be represented in different, complicated, and most beautiful kaleidoscopic patterns, very poor ideas of a few of which are shown in Fig. 104. At first we may get only shadowed figures, but with a little practice we soon get exquisite colour figures, with symmetrically-arranged whirling vortices as well; and if we sing a song, every change of note

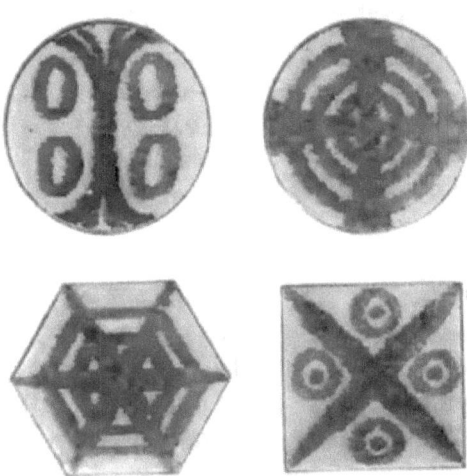

FIG. 104.—Phoneidoscope Effects.

will be optically represented on the screen by a corresponding figure. The sounds of instruments may be shown in the same way if sufficiently strong vibrations can be thrown into the mouthpiece; but the voice alone is amply sufficient.

This is one of the most magnificent optical experiments possible, and easily shown. By employing a mouthpiece with a vibrating plate or membrane, there is little risk (as with Tisley's form of apparatus) of bursting the film, and a good tough one will sometimes last a quarter of an hour.

In fact, when an unusually good solution has been obtained it should be reserved for these experiments with sound, ordinary samples being good enough for simpler phenomena. If the film breaks too soon, it shows either that the solution is not good, or that the discs have ragged edges, or are too dry: metal ones, which are best, hold the film more smoothly if heated and smeared with a *thin* coat of melted paraffin. If the mouthpiece cannot be conveniently made, the next best plan to avoid rapid bursting is to stretch the film on a ring, and adjust this about an inch clear away from the top of the vessel, A, so that currents of air may escape, and only true sound vibrations reach the film. Some solutions seem to do best in this way, which also avoids a slight rattle the mica diaphragm is apt to make; but on the whole I have found such a diaphragm mouthpiece safest and best.

103. **Fresnel's Mirrors.**—There are many other ways of producing interference between two rays of light than the use of thin films; but not all of them are capable of employment with the lantern, owing to the amount of light used not being sufficient to be visible when spread over a screen. Fresnel, for instance, letting a cone of light from a luminous point fall upon two mirrors very slightly inclined together from the same plane, formed interference fringes. The arrangement is shown in Fig. 105, where a pencil of rays from the sun is converged by the lens to the one point or line of emission we have already found necessary. A test-tube filled with water as a cylindrical lens answers perfectly well, or the line of light *reflected* in sunlight from such a tube filled with mercury will also answer. The diverging rays beyond the focus are then received upon the two mirrors, M m, m N, and if the inner edges or junction line of these be very slightly depressed, it is manifest the reflected rays will somewhat cross each other, and that light from both will appear on a portion of a piece of card held

as a screen at s. On this portion will be found dark and light, or coloured stripes, due to the interference of the waves. Two pieces of the *same* glass blacked on the back and laid on a piece of cloth on a flat board, the inner edge of one being depressed a little by the end of any pointed tool, will enable the student to perform this instructive experiment. If the light from the lantern be made to diverge from an extremely narrow slit, and the mirrors be arranged at several feet distance from it, the dark fringes may also be observed by the few who can gather round the apparatus; but on the ordinary screen the fringes are far too faint to be seen.

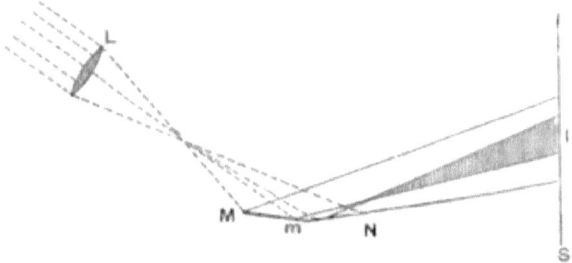

FIG. 105.—Fresnel's Mirrors.

104. **Fresnel's Prism.**—Fresnel also caused two beams to interfere, by interposing in the diverging cone a double prism of very small obliquity called an "interference prism," with the same result. Here, too, the light is too faint for a large screen when a single line or point of light is employed; but in this case, some approximation to the effect is possible. Prepare two or three sliders, the $4 \times 2\frac{1}{4}$ inch size, of blackened glass, and through the black, cut or scratch, over the field, perpendicular lines (Fig. 106) of uniform width and distance for each slide, but varying in these characters on different glasses. (Only one is really necessary, but it will be found that each screen distance

only shows good phenomena with its own gauge of bright lines, which must be found by trial.) Place in the optical stage and focus : then against the nozzle, N, hold, or fit in a tube which slides on it, the double prism, P, as in Fig. 107, which will give two images, whereof one set of slits will, more or less, overlap the other. At once, if the slits are the right gauge for the lenses and screen distance, we get colour;

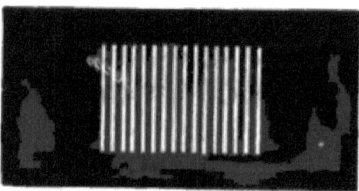

FIG. 106.

and though a little of this at each side of the screen may in some cases be due to the slight dispersion of the prisms added to that of the focusing lenses, it can readily be shown, by covering one half the prism, that nearly all the colour in the middle is not due to this cause, but to the interference of the two sets of waves. We may demonstrate

FIG. 107.—Fresnel's Prism.

this further by sliding over one half of the prism a thin piece of clear glass. By this, one set of waves is, of course, somewhat retarded ; and if we watch the screen attentively, we can see that the stripes of colour are a little *shifted* in consequence, as we move the plate backwards and forwards.[1]

[1] The private student will have no difficulty at all with this experiment. He has only to cut in a black card a set of slits, about one-

105. **Irregular Refraction.**—This latter phenomenon suggests another means of causing interference; viz., by irregular refraction, causing *retardation* of portions of the light. Such are the phenomena of what Dr. Young called "mixed" plates. Provide a few discs of plate glass about 2 inches diameter. Carefully clean two discs so that they show colour when pressed together, and then introduce between them a bit of butter or suet the size of a large pin's head, and some clean saliva or a drop of water; or the froth of white of egg well beaten up will do; or fine soap-lather. Work together with a circular movement, and gradually a film of mixed grease and water, or albumen and air, &c., will spread between the plates. Now it is plain the light which passes through a denser molecule of this mixture is more retarded than that which passes through the other; and hence we soon find, on looking through the film at a luminous point, there are beautiful halos of colour. Or if we place in the optical stage a black card in which a few small holes are made, on focusing the holes, covering the nozzle with another black card pierced with a $\frac{1}{8}$-inch hole, and holding a little in front of this (the exact distance must be found by trial) the "mixed plate," the images are surrounded by coloured halos, the colour of course depending upon the thickness of the film.

To the same cause must probably be referred the curious phenomena seen when an equal mixture of glycerine and spirits of turpentine is shaken together. What is known as diffraction will not account for it, since there is no approach to a spectrum; but it is completely accounted for by the

sixteenth **of an inch** in width and distance, and hold this at arm's length against a bright sky or the opal globe of a lamp, with the slits perpendicular. On now holding the double prism close to the eye, with the centre line over the pupil, he will at once see conspicuous colours. Such prisms an inch square cost from 2*s.* 6*d.* to 5*s.* each.

unequal retardation of the light by the two media in small molecules, which only mix mechanically. On looking through the mixture at any illuminated objects, they will be seen fringed with **colour**, the colour changing as the liquids again settle, till only **a coloured line is** seen in the fluid itself where the **two are in contact. I have** only tried the experiment thus, **but** by making a **small** closed tank with parallel **sides,** there is not the slightest doubt very beautiful phenomena **might be** projected on the screen. **There are** also minerals—such as what is called "iridescent agate"—which have this property. **I met with a "section"** some time ago, which was mounted over a hole in a blackened slip of wood. **It shows no** colour when interposed direct; but by turning the lantern **off at an angle,** condensing **the** light on the agate **also** at an angle, **and then focusing on the** screen "askew," as it were, all the **colours of the** spectrum may be produced according to the angles. This also **is probably a** case of irregular refraction. The mineral called Wulfenite often shows the same phenomena.

106. **Diffraction.**—One of Newton's two great difficulties about the Undulatory Theory was, that if it were true the ether-waves ought to bend round **the** edges of bodies into the shadow. It appears **strange now, that even** the few experiments **he made** in diffraction did not suggest to him that this is precisely **what** does happen. **If we hold** any opaque body, **such as a black** card, **in the rays from** a very small point **or line of** brilliant light, such as already described, we **find there is no sharp** shadow, but a series of coloured fringes due **to interference, both within** and without what should appear as the geometrical shadow. Dr. Young, who first pointed out the true character of these fringes, supposed **them due to the interference of** the direct **rays** with **those reflected at a great** obliquity from the **edges** of the body. This has been shown to be incorrect; **and** all the

fringes have been proved to occur from the interferences of the *secondary waves* shown in Fig. 75, when separated from the "grand wave" by the opaque body. They can be observed without any apparatus at all by receiving on a card a shadow from the planet Venus at its brightest, in an otherwise dark room.

But more beautiful phenomena are within our reach, the nature of one class of which is best seen from a simple experiment without the lantern, which should first be made by every student. Cut a slit $\frac{1}{8}$ inch wide in a black card, and hold it in front of a flame so as to be brightly illuminated. Blacken a bit of glass, and scratch with a needle a straight line on that. Hold the scratch close to the eye and look through it at the slit, held at arm's length, both being perpendicular. We see the slit in the centre, and on each side of it are a *series of spectra*, and we thus prove conclusively that the waves of light *do* bend round and spread out laterally from the second slit, becoming visible under these circumstances partly because all stronger light is cut off, and secondly, because we stop off the chief part of the main wave. (See Fig. 75). The spectra, of course, represent overlapping images of different colours, as we can see if we cover our first slit half with blue glass and half with red, we then get a series of red images *further apart* than the blue images, precisely as in previous experiments. The reason why we get the dark spaces between, and not one unbroken band of light, is that at certain intervals, which can be, and have been, exactly calculated, the waves from, say, one edge of the slit, interfere with the waves starting from the other edge, or other point, in it. From one edge the path to the eye is *longer*, and we have already learnt that retardation means extinction of certain colour-waves. (See § 110.)

107. **Gratings.**—This experiment theoretically ought to

be shown by the lantern, and it has been stated that it can be; but I have never been able to do so in this form, for the reason already given; the light passing through the second slit is not sufficient. Undoubtedly the spectra must be on the screen; but they are far too faint to be perceived. We must get "more light," and we are helped in this by the fact, that if we arrange a number of slits *exactly* at equal distances, their various interferences and correspondences all fall at regular intervals, depending partly on the width of the slits and partly on their distances apart. Such is what is termed a "grating," or series of very fine light and dark lines. If fine enough, such an assemblage of slits practically cuts out, at each point, all but one single wave-length, and so produces *pure* spectral colours only; whereas the other interference colours we have seen are mixtures of residual colours.

Nobert has ruled gratings with 3,000 and 6,000 lines to the inch, and photographic copies of the first of these are sold at a guinea, or less, and produce most beautiful phenomena. Placing a slit, say $\frac{1}{8}$ inch wide, in the optical stage, and focusing, we hold the grating just in front of the nozzle, with its lines parallel to the slit. There are at once projected on the screen, on each side of the central image, a most beautiful series of diffraction spectra (see F, Plate III.). Two similar gratings *crossed* give beautiful spectra of a small hole about $\frac{1}{8}$ inch diameter. There are not only perpendicular and horizontal spectra, but also diagonal ones, at all angles radiating from the centre, but fainter; these latter being of course due to the dispersion into fresh diffracted images, by the second grating, of every spectrum produced by the first. The finest effects are, however, produced by placing in the optical stage a symmetrical pattern of *several* small holes—say eight arranged in a square; but the pattern, size, and distance must be found by experiment and adapted to the gratings used and the screen distance. If properly adjusted, and with sufficiently brilliant light, most beautiful diffracted patterns will appear on the screen.[1] A circular grating scratched on glass gives beautiful circular rainbows when interposed in the path of the rays from a small aperture focused on the screen. Any kind of grating gives most brilliant effects if held close to the eye, and looked through at any small naked flame.

Another attractive method of observing diffraction phenomena is to cover the object-glass of any telescope —even one such as can be bought for 7s. 6d. will do— with caps of black card, in each of which is pricked or cut one or more very small holes of various sizes, arranged in different patterns. Or the cap may be of blackened glass, on which is scratched very small any regular curve or figure, such as a tiny circle **or** square. On looking through the telescope thus furnished at a bright star, beautiful diffraction figures will be **seen; or** if the telescope be directed to a small hole in a plate close in front of a good lime-light, the phenomena will be gorgeous beyond description.

But such apparatus is not needed to show diffraction. The two slits have already been described, and even the fingers will show fringes if *nearly* closed and looked through at a distant candle. With care and a screw-feed, any one may rule on smoked glass for himself, with a sharp needle, a grating of 120 lines to an inch, and even this will give very perceptible spectra. Wire-gauze is made 120 meshes to

---

[1] **Mr. C. J. Fox, F.R.M.S., first** showed me this beautiful experiment in the microscope, to which he had adapted it, and in which the effects, owing to better illumination, are far superior. For that instrument the apertures, pricked in a blackened circle of paper, are focused by a low-power objective, and the gratings mounted so as to rotate over the eyepiece. With the lantern, the parallel beam from even the lime-light needs to be condensed by an extra lens into the smallest area which will cover the apertures.

the inch, and sometimes 150 meshes; and this will give perceptible phenomena. So will a selected bit of the very finest cambric, such as is sold at 20*s.* per yard; but a good bit can only be found by experiment, when it should be mounted between two glass plates. Even more simple objects are at hand. Get a broad pheasant's body-feather (some birds' are better still, but this is easily got; most fowl feathers are too coarse), and mount it between two glass plates, blacking out all the space the feather does not cover. Focus on the screen a small hole in a black card, and interpose the feather; or look through it at any naked gas-light; again we get attractive diffraction phenomena. A very familiar example is found on looking through almost any railway carriage-window, at night, at one of the lamps. The dripping of the rain and dust, and the process which goes by the name of cleaning, combine to form tolerably perpendicular lines on the glass, and these draw out the luminous point into a long band. Usually this is only vaguely luminous,—the lines being irregular, various colours overlap and produce white—but on one or two occasions in the course of several years I have seen colour. Colour can also be seen on looking through glass breathed on so as to appear dull; but a better method is to dust through cambric on a glass some lycopodium powder. This will give fine rainbow halos when looked through at any luminous point, such as a distant candle; but its *lantern* effects are poor, the image appearing too bright and the halos too faint for satisfactory effects.

Most beautiful lantern projections may however be produced from perforated cards, such as can be bought for book-markers at a penny each. Several sizes are made, the smallest being about twenty-five holes to the inch: purchase one or two of each size, and blacken all but the smallest, which would fill up or choke. Cut circular discs to fit such standard wooden frames as already mentioned, and fix in such frames, by a spring circular wire, two of each size. First place in the optical stage one such slide (generally a medium gauge does best for this; but again, much depends on the screen distance). On focusing, we of course get images of the circular holes. Now gently *run in* the focusing-tube a little further, so that the focus is projected rather *beyond* the screen. The pencil of light from each aperture now interferes with its neighbours, and we get a more or less decidedly *coloured* pattern, which varies as the lens is moved in or out of the focusing-tube; at two or three **feet away** from the nozzle, the colours are pretty vivid. Much interest will be found in the various interference patterns **given even by** single cards **in** this way. But now add a *second* slide **in** front of the other. If the frames are $\frac{3}{8}$ of an inch thick, and the slides are inserted "the same way," the cards will be that distance apart, which, with a screen five feet off, is for most lenses a good distance; if a little more or less interval is necessary, one slide may be reversed. The pencils of light from the posterior holes are now further diffracted by the second set. **As** a rule, I have found the best effects at four or five feet distance, the *back* card being focused somewhat short of the screen, and from medium gauge cards; but fine effects are produced at that distance (which should not be exceeded with a lamp or burner) with most of the cards, or by a coarser one behind with a finer pattern in front. A little experiment will soon be repaid by most beautiful coloured tartan patterns, especially if the front card can be rotated in a frame such as is described later on for polariscope experiments; all produced by the interferences and diffractions of the small pencils of light. With the lime-light and a screen distance of ten or twelve feet, I have found the best effects from using both cards of the finest pattern, placed back to back, so as to be only $\frac{1}{8}$ inch apart. As one card is rotated, beautiful effects will be produced when the right focus is got; and in certain positions it will be found that a little alteration in *focus* only, appears to *rotate* large squares of the pattern in a most beautiful manner. But the arrangements are different with each objective and screen distance. The small white screen will be found very handy to show the varying effects at the shorter distances, and any experiments intended for exhibition should be carefully worked out at *exactly* the distances that **are** to be employed, **or** the effect may be found quite different from that intended. Any experimenter, however, may depend upon finding great pleasure in this direction. The private student may hold one card close to the eye, and the other at a little distance.

108. **Striated Surfaces.**—Lastly, if we get light *reflected* from very narrow lines or grooves, we ought also to expect interference. That we do obtain it, is shown by several beautiful experiments. Turn off the lantern parallel with the screen, and placing a perpendicular slit in the optical stage, focus it as *reflected* from a bit of plane glass held close **to the nozzle at** an angle of 45°. **Then** substitute for the plane glass, a *grating* held in the same position; the image **of the** slit is now flanked on each **side** by spectra, just as when transmitted light was used, only not quite so bright. **If** no grating be possessed, get a finely-coloured piece of mother-of-pearl polished plane flat, or one of those beautifully-coloured shells found in Jersey which can be bought for one shilling almost anywhere, or a finely-coloured pearl card-case will do; the main thing is to select a richly-coloured specimen. There is really no colour *in* these shells whatever: it is entirely due to the interferences of rays reflected **from** the countless little grooves which cover their surface, **as can** be proved by taking a good impression of one of the **best bits in black sealing-wax, when** the indented wax will

show the colours of the pearl. Adjust the **shell or piece of pearl** like the soap-film, condensing **all the light on it at 45°,** and focusing with the loose lens. The **coloured image** shows **a kind of glowing** "transparency" on **the screen, very** beautiful; **but as we** now gently **turn** the shell **so as to** change the **angles,** *the colours change,* showing **that they are not fixed, but due to interference.**

' Even a **peacock's** feather **gives beautiful phenomena** treated in the same way. I **have not been able to satisfy** myself that *all* the colour in this **is** due to **interference of**

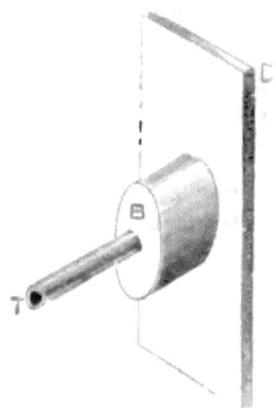

FIG. 108.—Tablet of Objects.

the reflected rays, but certainly **a** great deal **is; for if we** mount the end of one by two black ligatures sewn **through a** black **card, and treat it** like the pearl, **we shall find that at different angles we** get very **different colours,** passing from **deep purple to** brilliant green. This is a beautiful experiment, but requires the lime-light **at not** too great a screen distance. **For all** these latter experiments, the handiest plan **is to prepare a small** blackened tablet like Fig. 108; a thin **piece** of blackened **deal, D,** being glued on a boss, B, into which is driven a tube T, fitting in the sockets of our pillar-stands. The feather, or shell, or other article, can either be fastened direct to the tablet with two elastic bands, if solid and small enough, or the feather on the black card can be affixed by two blackened drawing-pins. The object can then be either rotated on a horizontal axis, or the angle of incidence and reflection rotated vertically by turning the pillar.

The metal buttons engraved with very fine lines, formerly made by Sir John Barton, then Master of the Mint, and still known as " Barton's buttons," are amongst the most splendid examples of the interference of reflected light. They are very scarce now, none having been made for thirty years; but a few are treasured here and there, and when the polish is well preserved, they glow with brilliant colours impossible to describe.[1] Any who possess even one, in good condition, may produce an exquisite screen projection with the lime-light by taking out all the front lenses, placing a $\frac{1}{4}$-inch aperture in black card in the ordinary lantern-stage, and allowing the beam from this to be reflected from the button to the screen. Or the naked lime-light, through a small aperture in a cap slipped over the empty condenser-cell, may be *focused* on the screen, which gives more light. Various small spectra will be seen arranged in a beautiful stellar pattern depending on that of the button. I have seen also very good effects from the finest cut that can be produced in the lathe from a pointed tool on a flat surface of bright metal.

[1] By the great kindness of Mr. John Barton, grandson of the inventor, I was placed in possession of a fine set of these buttons or " Iris ornaments." I learnt from him that the dividing engine, constructed for Sir John Barton by Messrs. Maudsley and Co., had been always in possession of the family since; but that since his father's death, no one for a long while could use it efficiently. After many trials, however, Mr. Robert Barton, when in Australia with it, succeeded in producing good results, and some specimens of his workmanship were shown at the

109. The Diffraction Spectrum.—In all these ways we have produced colour, or **dark** fringes, by the interference of different series of waves. It will be manifest that, as before, we have got all our colours **solely by suppressing** or quenching colour—that the colours are precisely of the **same** character **as** the dark fringes. It is equally manifest that since, as a rule, only one particular wave-length can be *completely* extinguished at any given point, the colours we see **are** not pure, but compound residuals, or made up of the residue of the spectrum. The only exception is in **the** case of gratings, which, by the orderly sequence of successive extinctions, cut out all, or nearly all, except **one** wave-length at a given point. A little thought **will show** that in these grating spectra we must therefore **get** the colours, **or** Fraunhofer lines which locate them, **in the** *real order and proportion of their wave-lengths*, **and not** as affected by **the** various **or anomalous** dispersions of refracting prisms. This **is so ; and the** fact **makes** grating spectra, though **less** brilliant **than** those given by **a** prism, owing **to** the large suppressions of light which produce them, especially valuable to the physicist.

This difference **in** the spectra, **and the** uniformity of diffraction spectra as compared with those produced by a prism, can easily be shown by experiment. Diffracting the pencil from a brilliant small aperture by a parallel grating, and further diffracting the *spectra* thus produced by a second similar grating held with its lines at right angles to those of the former, we **have** already seen (§ 107) that we get a most beautiful series of diagonal spectra. This, of course, must follow from considerations already discussed, and the greater distances apart of the red images

---

Melbourne Exhibition of 1880-81. There is therefore some hope that these beautiful objects may before long be again accessible to such as admire them.

than the blue. And all these diagonal spectra are perfectly *straight*. But if, instead of a second grating, we use a prism, with its refracting edge in the rectangular line, it is not so. We still disperse not only the central pencil of white light, but each **spectrum** produced by the grating; and, as before, the blue portions of the spectra are more refracted than the red. But the deflection is no longer *proportional*, but dependent on the special dispersion of the prism; and hence the refracted spectra now appear as parabolic curves, represented at G, Plate III.

110. **Measurement of Waves.**—It is plain that we have in the phenomena of interference, various means of

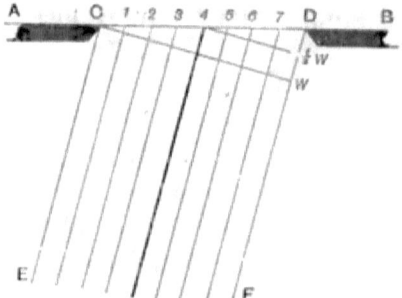

FIG. 109.—Nature of Diffraction.

measuring the *lengths of the waves* which produce any given colour. For many and obvious reasons such measurements are easiest taken with monochromatic light; and the simplest case is that of the light from a slit or point passing through a second slit (§ 106). Let A B be a highly magnified representation of the second **card, and** C D of the slit in it. The rays which pass perpendicularly through C D will none of them be retarded, and therefore produce on the retina or a screen an ordinary white image—the central white band. But as the card A B stops off the main wave-front (§ 62) every particle of ether in motion all across the width of the slit produces new secondary waves spreading right and left of the perpendicular direction: let us take any given inclination, C E, D F. Then drawing C w perpendicular to the course of the ray, we see that the waves from D have further to go than those from C by the distance D w. Assume that in this case D w is a wave-length of the colour employed, and number the supposed ether particles across the slit from unity onwards,—then we see that from 4, the centre particle, the waves are half a wave before those from D, and half a wave behind those from C. But still further, the rays from 1 will be half a wave before, or in complete discordance with those from 5, 2 with 6, and so on: every single ray finds another in complete discordance with it somewhere in the slit: obviously therefore *at that particular angle there must be a dark band*. Further to the right or left the relations will alter, and there will be a bright band, to be succeeded by other dark and bright bands. A general proof of the correctness of this reasoning is found in the fact, that plainly, according to the theory, the narrower the slit the greater ought to be the angular distance of any given band from the central image. This must be, because it will demand a greater obliquity to make the necessary difference of paths from the edges of a narrower slit. Experiment shows that this is the case.

Upon this hypothesis we can measure exactly the length of our wave, as in the diagram, Fig. 110. Draw the line A B to represent the card as before, with C D, the slit in it, and C F, D F, the direction of the first dark band from the centre of the field. With the radius C D describe a semicircle, and from C also draw C w as before, perpendicular to D F; then D w is one wave-length, and $x\ y$ is the angular value of the obliquity of the dark band. Inspection shows us at once that the angle x D y is necessarily equal to D C w; and as D w is for such a small distance practically coincident with a segment of the semicircle, **we only** have to take the proportion of 180° (the semicircle) to the angle D C *w* (the obliquity), **and** the same proportion **must** exist between **the** linear *length* of the semicircle and that **of the** wave. Schwerd found that with **a** slit of 1·35 mm. the angle D C *w* was 1' 38", the **ratio** of 180° to which is 648,000 **to** 98. The linear length of the semicircle is 4·248 **mm.** It follows that the length of the wave, of the colour employed **in** his experiment, is about the $\frac{1}{40,000}$ of an inch.[1]

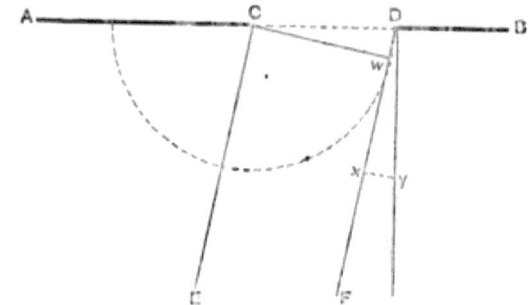

Fig. 110.—Measurement of wave-length.

We may reason in **the same** way from thin films. The thickness of **the** film **at the** first bright ring in Newton's **lenses,** with the **same colour employed by** Schwerd **in** the above experiment, **is** found to be about $\frac{1}{160,000}$ **of an inch.** This thickness **we know** has to be doubled to give the retardation, which gives us $\frac{1}{80,000}$ **of** an inch for a *retardation* that causes **a** bright **ring.** Here then is an apparent discrepancy ; **for** according to the calculation with our slit, $\frac{1}{80,000}$ **of an inch should** only be *half* a wave, and the ring ought to be *black*. **One** or **other of** the reflected rays is *half a*

---

[1] **The** phenomena of gratings are too complex to enter into here ; they are admirably elucidated in Müller-Pouillet's *Lehrbuch der Physik*.

REVERSAL OF PHASE.

wave out, or is in the contrary phase **to what** the mere retardation produces.

But on further reflection this apparent **contradiction** proves the truth of the theory, since we perceive it ought to be so. We have seen that waves involve periods and phases even more essentially than lengths; **and we** have to consider what happens when a wave is reflected from **a** *denser* or **a** *rarer* medium respectively. Take for illustration, as simplest, our first example of wave motion, in a set of ivory balls. Let **from A to B be** large balls, and B to C much smaller ones, and let them be united together by an elastic cord *a c* through the centre of all. Then if **we** roll a ball up against A, the wave of *compression* is **transmitted to** B. There it throws off the ball D, which in turn passes **on**

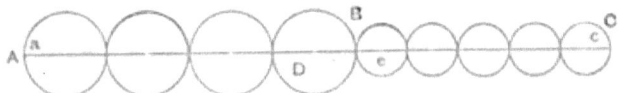

FIG. 111.—Dense and Rare Media.

the wave through the smaller balls. **But** the small balls are not enough to take up *all the motion* as the large ones did; they are driven off more freely, in strict analogy to **a** wave encountering a *rarer medium*. The last ball D has therefore motion to spare, and *follows* the first small ball *e*, but slower and more feebly. In this it is however checked by the elastic cord, which instantly pulls it back, or is rather pulled by it; and so the effect is really a *pull* upon the balls behind. In other words, the wave of compression is, *in the very moment and in the act* of reflection from the medium of less resistance, changed into a wave of extension. In other words again, it is converted into the opposite phase; or yet again, is thrown half a wave-length out of phase. If, however, **we reverse** the process, roll the loose ball against the smaller row, so as to impart the impulse to c, it is not so. The last small ball *e* finds a *greater* obstruction to its motion instead of a less, and cannot therefore have any tendency to fly off; the reflected wave remains a wave of compression, as it was at the moment of impact.

Thus we see that when light is reflected from two surfaces of a film, one of which is the surface of a denser and the other of a rarer medium, the reflection which is of the latter character must be thrown half a wave-length, or half a phase, which is the same thing, out of its order. In a soap film it is the second surface; in the film of air the first surface. The result is that the retardation which, judging by *thickness alone*, would have retarded the ray from the second surface half a wave-length, is altered another half wave-length forward or back (it does not matter which, the alteration to the *opposite* phase of vibration being the point,) and the two rays are brought into accordance, or give the bright ring. This measurement, therefore, now gives the same wave-length as the other.

This explanation also can be subjected to experiment. If we use a top lens of very low density, and an under plate of great density, we can introduce between them a film of fluid of some *intermediate* density, as oil of sassafras. Both reflections then take place from a *denser* medium, and the retardation alone should come into play, without any other alteration of phase. Experiment fulfils this expectation to the letter.[1] The apparent objection thus becomes another strong argument for the truth of the theory.

All the other phenomena of interference are similarly capable of calculation. And the most complicated phenomena have so far corresponded with calculation in the most minute particulars.

---

[1] This explanation is taken in substance from Sir John Herschel, his popular exposition of it being the best I am acquainted with.

111. **Size of Matter Molecules.**—I am not willing to conclude this chapter without some explanation of the manner in which the dimensions of these light-waves throw light upon the dimensions of the molecules of matter itself. We are forced to conceive of matter as consisting of detached molecules, or separate very small portions, on account of the enormous power of expansion when heated which matter possesses. Moreover, while water and alcohol expand as one to three in the liquid form for the same increase of temperature, in vapour they expand *in the same ratio*,[1] which we can only account for on the supposition that the molecules are now at enormously greater distances, so that their *special* action on one another has ceased, and they only obey the general laws of gases exposed to heat. We are therefore confronted with these detached molecules of matter; and some of the main questions physicists are now investigating, are, What are the probable sizes and other properties and relations of these molecules?

Now on the **first** of these questions the measurements we have just obtained **throw** very considerable light. First, the molecules must be considerably less than those waves in dimensions, or they would be at least partially visible with the powerful microscopes we now possess. Nobert's test-lines of 112,000 to the inch—half the dimensions of a blue wave—were resolved in America by Dr. Woodward. Moreover, in many transparent bodies at least, as the waves pass through amongst the molecules without being very sensibly destroyed or affected, this **is** another proof that such waves are large in proportion; **as** they plainly are not split up amongst them, but as it were *surge* grandly over them with little resistance. And yet, secondly, the molecules cannot be *infinitely* less relatively—or, shall **we** say, tremendously less; because if they were, it is easy to see that a difference in

[1] About $\frac{1}{491}$ for each rise of 1° Fahrenheit.

the waves of one half, (we may take red light as $\frac{1}{36,000}$ and violet as $\frac{1}{58,000}$ of an inch,) could not so profoundly affect the phenomena as it does. To use an illustration I have seen somewhere, sawdust would show no perceptible difference in its effects upon water waves 30 feet and 50 feet apart; but logs of wood probably would. All this is indefinite, and yet it does give us a notion of what physicists call the class, or *order*, of magnitudes involved; and from some other considerations of this kind which cannot be given here,[1] it has been argued with very great probability that the average distance from molecule to molecule can hardly be more than a thousandth part, and hardly less than one ten-thousandth part, of the average length of a wave.

It is a little singular that very analogous deductions may be drawn from the phenomena of a soap film. With a good solution, if a film is stretched upon a ring as in Fig. 96, and carefully observed, after a while coloured bands cease to form, and a large white patch appears, answering to the first bright ring. This we know (§ 110) to be a thickness of *one-fourth* of a wave-length. But after this comes a patch of very dark grey, often called black.[2] Now the peculiarity about this is, that the boundary edge is perfectly sharp, as if cut with scissors! It is not so with Newton's lenses, where the diminution of thickness is gradual; and the only conclusion is, that there is here some sudden change in the thickness and therefore physical constitution of the film. It cannot be *nearly* one-fourth of a wave-length, or a considerable portion of light would be reflected. It must be as compared with the wave-length practically *nil*, for either (1)

---

[1] Most of them are discussed towards the end of Tait's *Recent Advances in Physical Science* (Macmillan and Co.).

[2] It is often stated that this grey only comes in patches, and that the film almost immediately bursts. With solutions made as described on page 161, I have had half the film remain of the dark grey for hours.

the two reflecting surfaces are so close that the retardation is practically nothing, and discordance is produced solely by the half-wave difference of phase due to the denser medium; or (2) the film is so thin that the air on both sides is in optical contact and there is no reflection at all. Obviously the first supposition is the true one; and it is **very easy to** see that if the film exceeded in thickness *one-fortieth* **of a** wave-length, we must have some traces of colour. We have here, then, apparently **an** abrupt transition **from** $\frac{1}{100,000}$ of an inch to something almost certainly not much greater than $\frac{1}{1,000,000}$ of an inch; and it is difficult not to believe that this must be due to some peculiar change **in the** physical *plan*, or constitution of the film; which again must almost certainly be in fairly numerical relation with **the size of** the molecules. And as we know that the mechanical equivalent of the heat required to vaporise a grain of water would not be sufficient (according to the law of capillary attraction) to reduce it **to a** thickness of $\frac{1}{500,000,000}$ of an inch, and therefore at that thickness the molecules could no longer hold together, but would separate in vapour, we seem to have here two outside limits **between** which the size of **the** molecules, or rather **the** distances between their centres, *must* lie.[1]

Finally, however, we can hardly suppose that a film only one molecule thick would hold together at **all. We** must therefore multiply the lesser limit by some figure, at least **2**; and we shall be quite within the mark in estimating the molecules in the film's thickness at from **2 to 5**. And even this low figure brings the limits of measurement for the molecules of this form of matter as something like $\frac{1}{1,000,000}$ of an inch **for** the greatest possible distance between the

[1] Sir William Thomson believes that the molecules of gas cannot exceed $\frac{1}{1,000,000}$ of an inch; and Mr. Sorby estimates various molecules as probably from $\frac{1}{1,000,000}$ to $\frac{1}{100,000,000}$ of an inch.

centres, and $\frac{1}{250.000.000}$ to $\frac{1}{100.000.000}$ (according to the multiplier we assume) for the least distance.

It is remarkable that several other lines of investigation lead to similar conclusions; but they need not be mentioned here. Only the merest outlines of the optical argument have been given; but these will suffice to show how Light is still, in another sense, a Revealer of those minute elements which can never be seen by mortal eyes.

APPENDIX TO CHAPTER IX.
Diffraction in the Microscope.

The phenomena of diffraction described in this chapter have a very important bearing upon microscopical investigation, and especially upon the advantage of increased angular "aperture" in microscope objectives. That the increased angles obtainable by immersing object and objective in a fluid, instead of observing the object in air, gave marvellously increased powers of definition, had long been known; but so long as this was supposed to be due merely to greater illumination, or the collecting of a larger pencil of light from the object, it could not be satisfactorily accounted for. At length Professor Abbe pointed out the true nature of the advantage gained, and the matter was soon demonstrated by ingenious experiments devised by himself, Mr. Stephenson, and Mr. Frank Crisp, so well known as principal editor of the admirable *Journal of the Royal Microscopical Society*. The following brief explanation is condensed from Mr. Crisp's lucid summary of the subject in that journal for April, 1881, to which I am also indebted for the diagrams by which it is illustrated.

It will at once be understood, from the phenomena of "gratings" already investigated, that if between the reflecting mirror and the stage of the microscope we interpose a very small opening in the diaphragm, and on the stage lay a "grating" of ruled lines, on removing the eye-piece and looking down the tube we observe a series of images of the aperture like Fig. 112, all circular in homogeneous light, but the outer ones consisting of spectra in white light. The small pencil admitted through the diaphragm is "diffracted," **just as we**

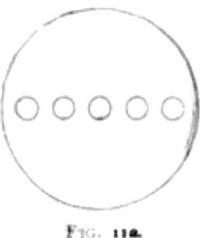

Fig. 112.

have already found. We next lay **upon the** stage a slide such as Fig. 113, consisting of, let us say, a circle containing both wide and narrow lines ruled on glass. Removing the eye-piece as before, **we** have of course, on looking down the

Fig. 113.   Fig. 114.

tube, the appearance presented in Fig. 114, the coarse lines giving diffraction spectra twice as close and numerous as those caused by the fine lines. The reason for this we have already seen (§ 110): the present point is, what influence these diffracted rays have upon the image, and it is here that the experiments just referred to are so important and interesting.

First of all, by a diaphragm at the back of the objective such as that in Fig. 115, let us cover up *all* the diffraction spectra, allowing only the direct, or central white pencil, to reach the conjugate focus, or image-point. On replacing the eye-piece, all the fine ruling has disappeared, leaving only

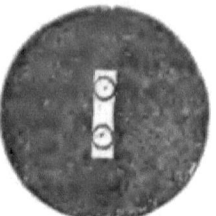

Fig. 115.

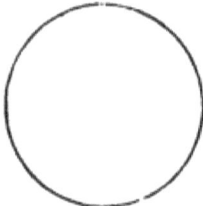

Fig. 116.

the general outline of the object, as in Fig. 116. By suppressing the diffracted rays, therefore, fine detail or "structure" of an object is *obliterated*.

Secondly, let us adjust behind the objective a diaphragm like Fig. 117, which allows all the lower spectra in Fig. 114

Fig. 117.

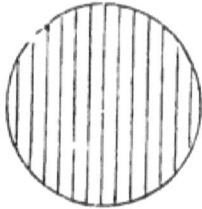

Fig. 118.

to pass to the image-point, but suppresses every alternate spectrum of the upper set, diffracted by the coarse lines. The image now appears as in Fig. 118, the upper set of lines to all appearance being identical with the lower set. Precisely in the same way, if we substitute a diaphragm like Fig. 119, stopping off yet another half of the alternate spectra, the lines are again apparently doubled, and we "see" Fig. 120, though the actual object remains the same. In these experiments therefore, while retaining the central pencil of light throughout, we have *created* apparent detail or structure in the object by suppressing certain of the spectra.

FIG. 119.

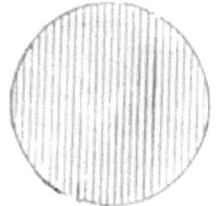

FIG. 120.

Still further, however, let us take a slide which when magnified resembles Fig. 121, or a "crossed grating." We get with this, from the small aperture, rectangular spectra somewhat like Fig. 122; but in addition there are

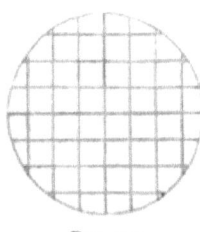

FIG. 121.

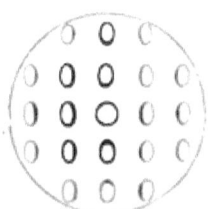

FIG. 122.

other *diagonal* spectra caused by the regularly recurring intervals diagonally across the squares. Constructing a diaphragm like Fig. 123, which allows only the central pencil and two of these diagonal spectra to pass, the vertical and horizontal lines of the object have vanished, to be replaced by Fig. 124. This experiment is troublesome, the diaphragm having to be prepared with extreme care; but the results, first deduced from theory, have been rigorously verified.

Now the microscopic student knows that many objects, by their minute and regularly recurring "structure," cannot fail to give, and do give, strong diffractive effects. The

Fig. 123.

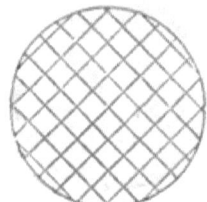
Fig. 124.

well-known *Pleurosigma angulatum* will serve as an example of the practical effect of the foregoing considerations. It gives three sets of diffractive spectra arranged as in Fig. 125.

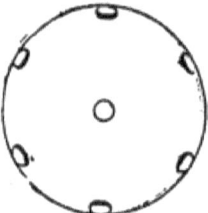
Fig. 125.

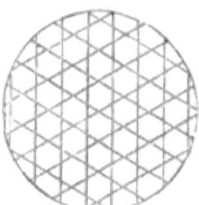
Fig. 126.

As each set is produced by something resembling lines at right angles to it, the three sets of lines in the object must be arranged *mainly* as in Fig. 126; but it will be obvious from what has gone before, that by selecting different sets of spectra, with or without the central beam, the apparent images will differ widely. It is also manifest that *all* these images cannot represent the *true* structure. If, however, we

are pretty sure that we have all the characteristic spectra, and
their position and relative intensity can be calculated, then
the resultant image can also be calculated; and *so far* as all
the spectra are included it will represent the real object.
Mr. Stephenson [1] records an extraordinary instance of such
calculation in the case of this identical *P. angulatum*. A
mathematical student who had never seen a diatom, taking
the spectra alone (as roughly shown in Fig. 125) worked out
from calculation the drawing given in Fig. 127 as the result.
Now the *small* markings between the hexagons had never
been seen in *P. angulatum* by anybody. But on Mr.

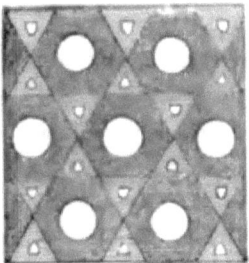

FIG. 127.

Stephenson making special examination of a valve, stopping out the central pencil so that its superior illuminating effects might not overpower the others, these small markings were found actually to exist, though they were so faint as to have eluded all observation until mathematical calculation from their spectra had shown that they *must* be there. Light was once more, even in the microscope, by its physical deportment, a Revealer of what the microscope had, up to that date, failed to see.

The general conclusion is, therefore, that we can have no *true* image of an object whose structure is sufficiently fine to

---

[1] *Journal of the Royal Microscopical Society*, vol. i., 1878, p. 186.

give diffractive effects, unless *all* the diffracted rays, or rather perhaps all the truly characteristic sets of spectra, are collected; and the image will more or less resemble the object, in proportion as the spectra are all collected, or at least sufficient of these *characteristic* spectra. As we have found before (§ 110) that the finer the grating, the more widely deflected are the diffracted spectra, we can now readily understand how, as regards minute structure especially, collection of the widest possible angular field of rays from the object is a point of the utmost importance for correct *delineation*, quite irrespective of greater *illumination;* and it is in this respect that immersion objectives have such an enormous advantage.

Of course these considerations only apply to structure of a certain degree of minuteness. With more coarseness, all the diffracted spectra which are visible may be collected by a moderate angle. But when we reach a certain fineness, it will be seen that the image in a microscope of small angular aperture can be no true representation of the object at all, but is due to peculiar selective conditions. This may be well shown by an experiment with *Amphipleura pellucida*. With a homogeneous-immersion objective of large aperture, focus the object under an illumination so oblique as to show up all the lines clearly. Then remove the eye-piece as in previous experiments; and placing the eye at the conjugate focus or image-point of the objective, the direct beam will of course emerge obliquely as a bright spot; while on the other side of the field, and *close to its margin*, will be seen a faint bluish light, the inner portion of the *first diffraction spectrum*. (Fig. 128.) Only a portion of *one* spectrum, observe; and that so near the margin that it must be lost with any objective of much less angle. If now a small bit of paper be adjusted on the back lens of the objective so as to stop this blue light and no more, the illumination is diminished by an almost infinitesimal portion, and the diatom is still visible, apparently as brightly illuminated as before. But the

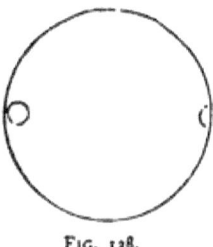

FIG. 128.

characteristic *striation*, which caused, and was therefore imaged by, the diffracted light, is gone, just in the same manner as was demonstrated in Figs. 115 and 116.

NOTE ON THE COLOURS OF THICK PLATES.

Nothing has been said in the text on the interference colours of *thick plates*; the phenomena being as a rule difficult to observe and unsuitable for projection. One exception may perhaps be made in an apparatus known as *Delezenne's Analyser*, founded on an experiment first made with two thick silvered mirrors by M. Jamin. If two truly parallel plates of glass are arranged thus, \\ , so that rays from the top of the page are reflected horizontally from the left-hand glass to the right-hand one, and thence to the eye below; so long as the glasses are truly parallel with each other there is no interference. But if either of them be rotated round an axis parallel to the side of the page, slightly to either side of this position, beautiful interference bands appear precisely similar to Savart's bands (§ 179). If the glasses are backed with black card to exclude extraneous light, and near enough to prevent direct rays passing between them to the eye, the phenomena are conspicuous against any bright surface, and readily projected upon a screen. Usually the glasses are mounted, one inside each of the two flat sides of a round brass box like a collar-box, one-half of which rotates in the other half; apertures being cut in the flat sides for the rays to enter and emerge at the interval between the glasses. Such an apparatus, about the size of a large watch, can be obtained of Mr. Tisley: but it is to be regretted that it was allowed to be described as if original before the London Physical Society in 1878, when it was simply a copy of Delezenne's apparatus constructed many years before.

CHAPTER X.

DOUBLE REFRACTION AND POLARISATION.

Double Refraction—Huyghens' Experiment of Reduplicating Images—Polarisation—Polarisers, Analysers, and Polariscopes—Phenomena of Tourmalines—Polarisation by Reflection and Refraction—What Polarisation implies—Analysis of Polarisation by Reflection and Refraction—The Polarising Angle—Analysis of Polarisation by Double Refraction—Extraordinary and Ordinary Wave-Shells in Doubly-Refracting Crystals—Action of the Tourmalines.

In dealing with light hitherto, we have found it reflected, or otherwise behaving according to certain uniform laws, which were not affected by the position or direction of the ray. We have now to examine phenomena in which that is not the case.

112. **Double Refraction.**—Place in the optical stage the smallest aperture that will show a bright spot upon the screen, and in front of the nozzle hold a piece three or four inches long of the clear mineral called Iceland spar, which crystallises in the form of a rhombohedron, as shown in Fig. 129. There appear perceptibly *two* images, separated by a slight interval; and if the spar is turned round pretty equally, it is seen that one image rotates round the other. It is plain that the single pencil from the lantern A B (Fig. 130) is, in passing through the spar, divided into two, B C and B D, and equally so that if one of these rays is refracted in the plane of incidence, and according to the law of sines, the other in some positions

CHAP. X.]   DOUBLE REFRACTION.   209

cannot be.   It is equally obvious that somehow or other the indices of refraction (§ 30) must differ in the two rays.

Large pieces of spar are clumsy, however, and give little separation, as both rays resume parallelism (C E, D F) when

FIG. 129.—Iceland Spar or Calcite.

they emerge from it. But as we have two indices of refraction in certain positions, it is evident that if we cut a prism of the spar with its refracting edge in the right position, the angular deviation will continue after the two rays emerge

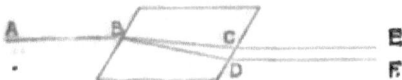

FIG. 130.—Double Refraction.

and the separation increase with the distance. The dispersion of the prism is easily achromatised, or nearly so, with a contrary prism of glass,[1] and a small bit of spar thus treated

[1] A better plan is to make both prisms of spar, cut in two directions at right angles with each other, on a plan devised by Dr. Wollaston. This gives better chromatic correction, and doubles the separation of the images.

P

gives us a wide separation. We shall suppose *two* such prisms, A and B (Fig. 131), mounted in cork, and so fitted that the brass tube containing the first rotates on the nozzle at N, while the second, B, rotates in the first, with a space, S, between them, through which, by slits in the sides of the mounting tube, a slide an inch wide can be inserted.

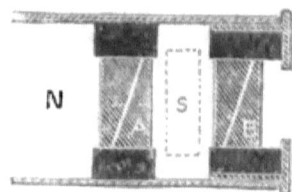

FIG. 131.—Huyghens' Apparatus for the Lantern.

Two double-image prisms thus fitted are usually called a "Huyghens' apparatus."

113. **Huyghens' Experiment.**—We can now use a larger aperture. Place one, $\frac{1}{4}$ inch in diameter, in the optical stage, and focus the image: on placing one prism on the nozzle

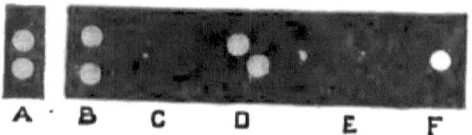

FIG. 132.—Huyghens' Experiment.

and rotating it, one image is seen to revolve round the other, but no difference in brightness or otherwise is observed. Let A (Fig. 132) represent these two original images. Add the second prism in front of the first, however; and keeping the first still (say with its two images perpendicularly disposed), rotate the other in front of it. Starting with both prisms in the same position, two images still appear, only at double the distance, B, showing that each ray from the first prism suffers no further division, but is only further bent in passing through the second. But directly the front prism is rotated in the least degree, *four* images appear; each pair being of unequal brightness, however, until one-eighth of a revolution has been made, or the second prism is at an angle of 45° with the first; all four are then equal, as at C. Proceeding, what were the faintest images become the brightest, and *vice versâ*, until when a quarter of a revolution is reached there are again but *two* equal images, this time, however, placed at an angle of 45° from the perpendicular on the screen, D. Still proceeding, the same stages are gone through reversed; but on reaching the half revolution, if both prisms are of equal separating power, there is but *one* image, F, into which all four have merged. The successive phenomena are represented in Fig. 132, and are of course reversed through a further half revolution back to the first position.

The student will do well to examine these phenomena more in detail, if possible, which he can easily do without any other apparatus than two small rhombs the size of Fig. 133. They can either be laid on a sheet of paper with one round black spot, or held against the window over a hole in a black card the same size. Let the under rhomb be kept in the same position as shown by the white figure, and the other rotated over it as shown by the shaded figure. It will first of all be seen that with the single rhomb, if so cut that the sides are of equal length, the line joining the two images is always parallel to the short diagonal. The same will also be noticed of the reduplication of the images; and the details of the diagram will enable the successive modifications to be accurately traced, and show all that takes place, the *white* spots showing a total extinction of the image.

114. **Polarisation of Light.**—It is manifest that the pencils of light which have passed through the first piece or prism of Iceland spar differ remarkably in *some* way from common light; and that the difference essentially consists in this: that they behave differently according to which of their *sides* are presented to certain sides of the second prism. We

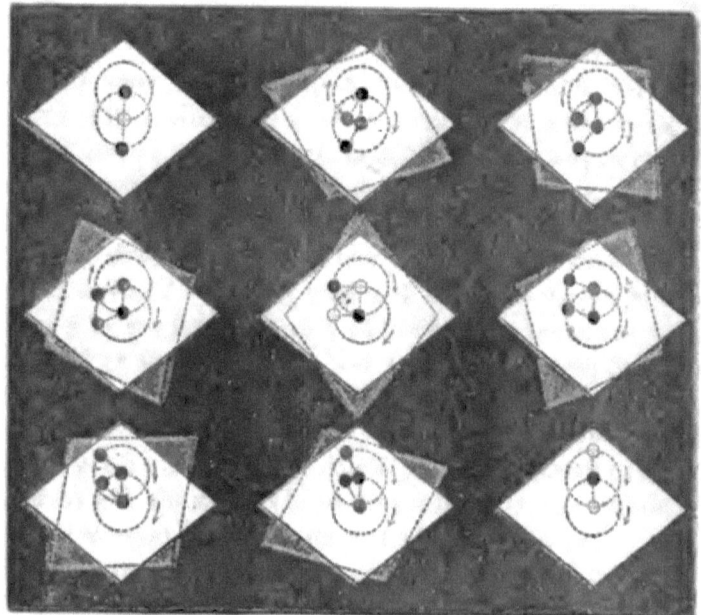

FIG. 133.—Analysis of Huyghens' Experiment.

have here an obvious analogy to the "polarity" of magnets and currents of electricity, which, though not strictly accurate, is sufficient to justify the term of "polarised" light.

115. **Polariser and Analyser.**—And the analogy goes further: as we cannot detect magnetic polarity until we bring to our presumed magnet some other magnetic or

diamagnetic substance by whose attraction or repulsion we detect the magnetism; so here we could not *detect* any "polarisation" in our two pencils of light, until we subjected them to a *second* process similar to the first. This law holds good throughout the subject. A great deal of reflected and other light around us is, as we shall immediately find, really polarised; but we cannot detect it to be so without subjecting it to some second process, which of itself would polarise it were it not polarised already. If it is, such a further process at once *reveals* the already existing polarisation, and the apparatus so used **is** then called an "analyser." A polariser and analyser together form a "polariscope." Any one of the methods which are capable of polarising light may be used equally **to** analyse light when polarised, whether it be the *same* process as polarised **it** or not being a matter of complete indifference **beyond** the convenience of the operator. These methods are several, and we have now further to experiment with them.

116. **Phenomena of Tourmalines.**—There is another doubly refracting crystal called tourmaline, some colours of which, when cut in slices parallel with the axis, have the property of rapidly absorbing, or being almost opaque to, one of the two pencils produced. Hence we greatly simplify the phenomena.[1] As one ray only passes through, which is the colour of the crystal, if we focus the slice upon the screen, we see nothing remarkable about it. Obtain, however, two

---

[1] Two mistaken statements **are** often made about tourmalines. One is, that green ones are good polarisers. Some few are, but many are not, and far the best colours are **the** various shades of browns, some of which are a very pure purplish grey, and very little change the colour of objects seen through them. The other error is that tourmalines "polarise by absorption." All that the absorption does is to take up or stop one of the *already-polarised* rays due to double refraction; for if a very thin wedge be ground, at the thinnest edge both images can be distinguished.

slices of tourmaline, of such sizes and shapes that one can be seen distinctly over the other. Let one be mounted in one of the 4 inches by $2\frac{1}{2}$ inches wooden frames, and the other on a loose disc of glass, which can be secured in a metal circle by a spring, and rotated by a pinion and circular rack.[1] Place both in the optical stage, parallel with each other, and focus; then rotate the front one: the successive appearances are as in Fig. 134. When parallel, A, there is simply a rather deeper colour from the double thickness; when the movable one is rotated 45°, as at B, a considerable portion of light is stopped where both are superposed; when at right angles, C, no light whatever can get through—the screen there is black.

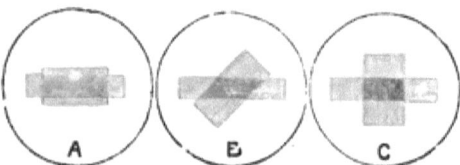

FIG. 134.—Two Tourmalines.

It is plain we have here the same phenomena as before, only simplified by the absorption of one of the two rays. To prove it, we remove the fixed tourmaline, leaving only the rotating one in the stage, and placing with it a circular aperture in a plate or card, just large enough to encircle the tourmaline. On the nozzle of the objective we place one only of the double-image prisms, which if of wide angle will quite separate the two circles of light with the tourmaline image in the centre; if the separation is not sufficient for this, remove the second lens from the objective,

[1] Such rotating frames, of the standard 4 inches by $2\frac{1}{4}$ inches size, can be purchased for a few shillings of any good London optician, and at least one is indispensable for many experiments.

and insert at C, Fig. 1, a lengthening tube or adapter, about 2¼ inches long, which, by reducing the size of the discs, but not their distance apart, will "clear" **them on the** screen. Adjust the prism **so** that the images stand horizontally, and the tourmaline stands vertically. One image transmits the light, the other is completely black (A, Fig. 135). Now **rotate** the tourmaline till it stands horizontally; the light image gradually becomes black, and *vice versâ*, B, whilst in passing through the angle of 45°, both are alike and semi-opaque. Rotating next the prism, while the tourmaline is stationary, the same alternations are repeated. **It is** perfectly clear that the tourmaline gives **us in a** single ray, precisely what the Iceland **spar gave us in** two rays.

FIG. 135.—Tourmaline and Double-Image Prism.

117. **Polarisation by Reflection and Refraction.**— In 1808 it was discovered by Malus that reflection from glass at certain angles gives the very same "polar" phenomena; and a few years later it was discovered that the refracted ray which passed through the glass had the same property. On a piece of board, B D (Fig. **136), as** base, glue or screw two triangular side pieces, B C D, and fix between the hypothenuse edges of these, B C, ten or twelve plates of thin crown or plate glass, so that the angle C B D is about 34°. It is evident that when laid on **a** table stand in front of the objective, a beam, E F, **from** the lantern will be partly reflected towards the ceiling as F G, and partly refracted and transmitted to the screen **as** F H. Adjust it thus: remove the double-image prism, but leave the rotating tourmaline in the stage. On

rotating the tourmaline, it will be found that **when** this is horizontal, the reflected image on the ceiling is bright, and when the tourmaline is vertical, black; and on looking at the screen we see **that** these effects, by the transmitted ray, **are** precisely reversed. And if **we** place an aperture in the stage **without** the tourmaline, and on the nozzle the double-image prism, **we, of** course, **find on** screen and ceiling reciprocally light and dark **images** of the aperture, **which** change to the opposite character **as the** prism is rotated. **Next** lay the bundle on its triangular side, and every image is precisely **reversed.** All through **we** have found similar phenomena; **and if we use** the pile of glass *first* (either as reflector or transmitter) and another pile of

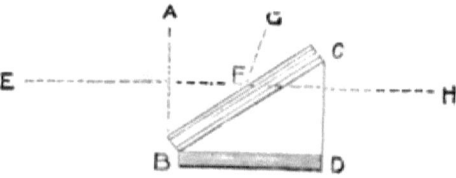

FIG. 136.—Glass Pile.

glass, **or any of** the other apparatus, after it, it is still the same; **the beam** of light, when "polarised" by any one of **these methods,** behaves **in opposite** ways when "analysed" by any one of the methods, in positions at right angles with **each** other.

A single plate **of** glass **is sufficient, when adjusted at the exact polarising angle, to polarise all the light** that is reflected; and an **equal quantity of polarised light** is also transmitted through the plate. **But this** quantity being small, and in the case of the transmitted beam overpowered **by the larger quantity** of common light also transmitted, it is **usual to employ a** pile **of at least** a dozen plates. **Even** this **does not nearly polarise all** transmitted light, but sufficiently increases the quantity of reflected light. Owing to inequalities in plates of glass, for **accurate** experiments it is sometimes necessary, when reflected light is employed, to employ only one plate of perfectly flat glass blackened at the back. It will further be speedily discovered that at other than the angle of complete polarisation, a somewhat *less* quantity of light is polarised.

118. **Explanation of Polarisation.**—The phenomena of double refraction and polarisation puzzled Huyghens and Newton, for opposite reasons. Newton's notion of alternate "fits" could be made to account in some measure for polarisation, but not for double refraction. Huyghens could not account for polarisation, but easily accounted for double refraction **on** the Undulatory Theory, even as then understood, when **it was** supposed **the ether vibrations** resembled those of **sound-waves,** or were propagated **in** the direction of the ray. **We have** seen that the retardation which causes refraction is caused .**by** *greater density* or *less elasticity* in the refracting body; and Huyghens had only to suppose **a** doubly-refracting crystal was less elastic in some directions than in others, to account for it, provided only the ether vibrations also were affected by these differences in the structure of matter. This, we have found from numerous experiments, is probable. But the theory as then understood failed to account for polarisation, which finally occasioned Young and Fresnel's great conception of *transverse* vibrations, by which everything is simply and perfectly accounted for.[1] Let it be supposed that common

[1] Familiar as the conception is to us now, it **is** difficult to realise what a profound and tremendous revolution in scientific opinion it was when first promulgated by Young and Fresnel in 1816 and 1817. To endow such a rare and subtle fluid as the ether with the most distinguishing property of a *solid*, was such a stupendous overturn of all previous notions about the Undulatory Theory, that Arago, who had up to that time shared and endorsed Fresnel's previous memoirs, shrank

light consists of vibrations in all azimuths, but all perpendicular to the path of the ray, as at A, Fig. 137. Whether vibration takes **place** in different azimuths simultaneously, or in succession, **is not** positively known, and does not affect the reasoning.[1] **Such a ray must** behave indifferently **as to its sides, whatever it meets with in** its path. But let all these azimuths **of vibration be "resolved"** into *two* planes

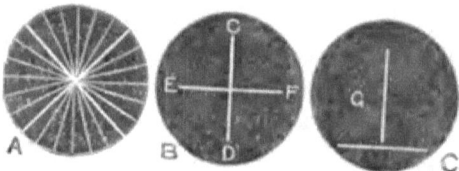

FIG. 137.—Nature of Polarisation.

at right angles to each other, as at B, where the top and bottom quadrants of A are supposed to be mainly resolved into C D,[2] and the others into E F; and suppose we can

from such a step, and left Fresnel to bear the brunt of it alone. Fresnel related, in 1821, that he himself hesitated to adopt it for a while, and states (see Whewell's *History of the Inductive Sciences*, vol. ii., p. 417) how "Mr. Young, **more bold in his conjectures, and less confiding** in the views of geometers, published it before me, though perhaps he thought **it after me.**" Whewell goes on to relate, from information given him personally by Arago, how when Fresnel **had** pointed out that **transverse** vibrations were the only possible way of translating the facts of polarisation into the Undulatory Theory, the **elder Frenchman "protested that he had not courage** to publish such **a conception, and accordingly the** second part **of the** Memoir was **published in Fresnel's** name alone." And **yet, when Arago** thus shrank **from the new** theory, he had received **also a letter from** Dr. **Young,** dated January 12, 1817, in which the **same idea** was suggested **for the same reasons!** Facts like these **should not be lost sight of, for the** sake of the instruction **they convey to the scientific student.**

[1] See Appendix to this Chapter.

[2] It results from very elementary principles of mechanics that **if C D** and E F **are the only possible directions of** vibration in a body, all vibrations in any other azimuth than one of the **two must be** resolved into

obtain either plane *separately;* for instance, the spar separates them somewhat as at C. **Very** plainly, such a single plane of vibration, on meeting reflecting **and refract**ing surfaces, must behave very differently according as **the** *ends* or *whole course* of each path **of** vibration come **in** contact with the reflecting or refracting surface.

119. **Analysis of Polarisation by Reflection and Refraction.**—It **is** even easy to see that reflection **and** refraction, of themselves, must **tend** towards this state of things. For tracing a ray vibrating in all azimuths, to an in**clined** surface, it is natural to suppose those vibrations which **come** in contact with it, or "kiss" **it, as** it were, along their whole path, should be reflected in their integrity, while others must be seriously affected. A simple experiment confirms this supposition. It is supposed throughout **the** Undulatory Theory that we are dealing with actual physical realities—with actual physical atoms, vibrating with inconceivable rapidity in **definite paths.** Now, any solid small **body thus** moving in an **orbit,** with sufficient **rapidity,** produces **on our** sense of touch **and** in many other mechanical respects (and we are dealing with strictly mechanical effects here) the same effect as a solid *occupying the dimensions of its path.* If, for instance, the driver-stud in the driver-chuck of a small lathe can be rotated with sufficient rapidity, it is in many respects mechanically equivalent to a solid *cylinder* bounded by the circumference of its circle **of** revolution. In the same way, a small sphere vibrating rapidly enough in a path perfectly straight, would produce effects similar to those of a small rod equal to it **in** diameter and of length equal to its path. We may thus conceive of the front of a ray of light,

both in varying proportions according to the angles. Near C D nearly all the motion **is** resolved into C D, and only a small portion into E F; while the vibrations in azimuths of 45° will be resolved equally into C D and **E F.**

supposing there are transverse rectilinear vibrations in all azimuths, as equivalent to a number of rods crossing the axis of the ray in all azimuths as A, Fig. 137. Now let us consider these various rods, preserving their rectangular relation to the ray, obliquely projected against the surface of some retarding medium, such as we have considered refracting media to be.

We can predict the result theoretically; but we can, as regards *single* rods, representing single azimuths of vibration, readily subject the matter to experiment by the simple instrument shown in Fig. 138. Let there be hinged on a

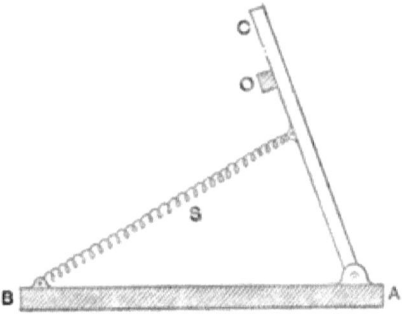

FIG. 138.—Catapult.

base-board, B A, at a pivot, a light frame C, pulled towards B by a strong spiral spring, S. On pulling it back and releasing it, the part C will act as a kind of catapult, and project objects laid upon it with considerable velocity at any inclination with the horizon, determined by a stop or obstacle, o. Provide a few smooth rods of wood, as accurately circular as possible, of different diameters and weight (this is simply because the best rods for each projecting apparatus must be found by trial: as a rule rather heavy wood is best), and placing the instrument in front of a shallow bath of water, project the rods obliquely against the water. Let a horizontal rod be first so projected. so that the whole length strikes the water at the same moment. With a little practice it will be found that when the rod is

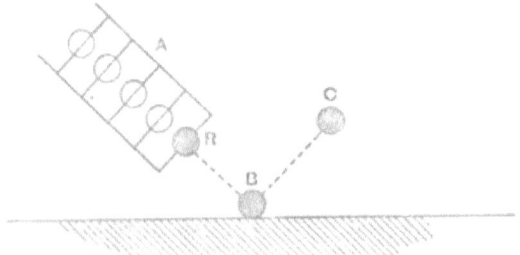

FIG. 139.—Reflected Rod.

projected pretty accurately, it is *reflected* from the water, as in Fig. 139, as boys play "ducks and drakes," or as a shot ricochets on the surface of the sea.[1] Now let A (Fig. 139)

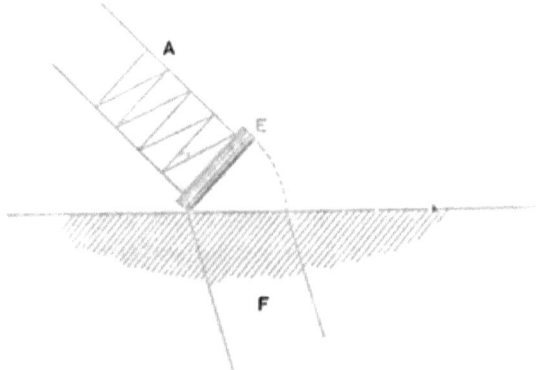

FIG. 140.—Refracted Rod.

be the wave-lengths in a ray of light, the rod R will represent those vibrations which reach the retarding medium

[1] The rods must strike the surface of the water at a considerably less angle than in Fig. 139.

as at B; and they are *reflected* to C. Lay next a rod *vertically* on the face of the catapult, so that the *end* strikes the surface of the water first, as at E (Fig. 140). Such a rod represents mechanically the vibration at right angles to the former; and it will be found it is no longer reflected, but swung round in some such direction as F.

But this is not all; for attentive observation, even with this rough and simple apparatus, will soon show that, in rods placed in intermediate positions, there is a sensible, visible tendency on meeting the water to *swing round* into one or other of the two positions we have examined. It follows that if we could project such a rod, without losing its energy, against surface after surface, it would gradually be brought into either one or other of these two rectangular positions. And this precisely accounts for what was long a difficulty, viz., the *gradually increasing* body of polarised light as common light is reflected from, or transmitted through, a greater number of successive plates of glass.

120. The **Polarising Angle.**—Lastly, a very elementary knowledge of geometrical mechanics necessitates the clear perception that, according to this purely mechanical method of analysis, the amount of reflected and refracted light polarised in two rectangular planes, assuming the original beam to contain equal proportions of all azimuths, must be equal; and further, that the most favourable position for the operative surface, or that which must give from any surface the largest quantity of each kind of light, must be at an angle of 45°, at which angle only an equal number of azimuths are affected in each of the two directions, and at which only the refracted ray is at right angles to the reflected ray. Now, at first sight, this seems to be contradicted by the fact that the polarising angle of glass is not 45°, but about 56°, and that it varies with every transparent substance. But when examined this difficulty disappears; for Sir David

THE POLARISING ANGLE.

Brewster discovered in 1815 the beautiful law shown in Fig. 141. Reference to the diagram of sines (Fig. 35) will show on mere inspection, that there must be a given angle of incidence, at which the ray, I R (Fig. 141), *reflected* from the refracting surface, at the same angle with the normal, N, as the incident ray, s, is at right angles with the refracted ray, I r. That angle in every case is the *angle of polarisation*. With glass it is about 56° 35′, but must of course depend upon the refractive index of the glass. Moreover, Sir David Brewster

FIG. 141.—Angle of Polarisation.

has also shown, by mathematical reasoning quite independent of the considerations here advanced, that there is every reason to believe incident rays are partially subjected to the deflection of the refracting substance before reflection, and that therefore the *real* polarising angle is in all cases 45°.[1]

[1] Sir David Brewster advances the following propositions: "When the refractive power of any body is infinitely small, its polarising angle will be 45°." "If light were polarised simply by the action of the reflecting force, the polarising angle would be 45°." "When a ray of light is incident at the polarising angle upon any substance whatever, it

It is easily seen that, as the refractive index of glass or any other medium differs for different colours, polarisation by reflection can never be quite perfect for *all* the colours of white light at any one angle. The imperfection is, however, so small, that it may be neglected for all but very highly refractive substances, or very delicate experiments.

121. Polarisation by Double Refraction.—Having determined the direction of the vibrations [1] in a "plane of polarisation," we can now examine the phenomena of a

receives such a change in its direction by the action of the refracting force, that the real angle at which it is reflected is 45°." "The real angle of polarisation is 45°, the effect of the refractive force being merely to bend the ray of light so as to make it suffer reflection at this particular angle." (See *Phil. Trans.* 1815, Part I.) The reasoning in this paper does not seem to me to have received quite the attention it deserves. It is the more remarkable because the "father of modern experimental optics," as Professor Stokes justly called him, was to almost the last, if not quite so, a disbeliever in the Undulatory Theory, and was not guided by any considerations connected with it. If his conclusions, and the mechanical analogy above advanced, be received, the supposed difficulty about the angle of polarisation by small particles (see § 187) entirely disappears.

[1] It is usually stated in works upon this subject, that it is an open question whether reflected vibrations are parallel (as here affirmed) to the reflecting surface, or in the plane of reflection; though it is usually admitted that the preponderance of argument is in the direction here adopted. The methods of experimental demonstration and purely mechanical reasoning here employed, leave, in my opinion, no doubt whatever upon the subject, and are, I think, more conclusive than even Professor Stokes's experiment—generally considered *almost* conclusive (see *Cambridge Trans.* 1850)—upon the effects of a fine diffraction grating on the plane of polarisation. It is to be observed that almost, if not quite every argument for the contrary view is based upon purely mathematical reasoning; and it is notorious that in purely mathematical questions opposite assumptions often give correct results, as in Newton's assumption that the velocity of a refracted ray in a denser medium was accelerated. The question, it is submitted, ought to be argued, and can only be decided, as one of actual physical mechanism, and not of pure mathematics.

doubly-refracting crystal. It has been assumed that double refraction is due to an *inequality of elasticity* in different directions. Is there then this inequality? Experiment shows us not only that there is, but that the physical properties of a crystal in this respect stand in fixed and invariable relation to its optical properties as tested by experiment; and that both these again have a fixed relation to its form. Some crystals are symmetrical in all directions, as the cube; if heated, they conduct heat equally in all directions; and with variations in temperature they expand equally in each direction. Now such crystals may safely be assumed to be

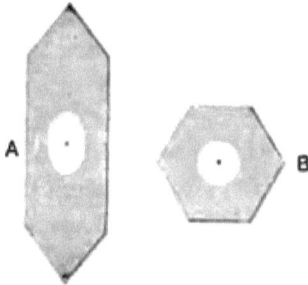

FIG. 142.—Quartz Plates.

equally elastic, and in a free or natural condition they have *no double refraction.* But if now we take a crystal of quartz, which crystallises in six-sided prisms with pyramidal ends, it is manifest on inspection that it is not symmetrical in all directions, but only round one axis, that of the prism. Take a slice cut parallel to this axis, A, and pierce it with a hole into which we can introduce a wire heated by an electric current or otherwise. Coat the plate with a film of wax, and introduce the heated wire; the wax will gradually melt around it, and it will soon be seen that the melted surface is an ellipse; in other words, the heat is conducted more rapidly along the crystallographic axis than **across** it. Doing the same with a slice cut *across* the axis, B, the melted area is now a circle; showing that conduction is equal in all radial directions round the axis.[1]

It has further **been** discovered by experiment, especially by Professor Mitscherlich, that **when** a crystal **of** this description **is** heated or cooled it expands or contracts *unequally* in different directions. Almost, if not quite invariably, a **moderate** heat expands the shorter axis of the perfect crystal **more** than **the others, or** brings the **crystal *nearer* to** the **form of a cube,** or other shape which in its most perfect and simple form can be inscribed in a sphere (this is the simplest general test of the nature of a non-doubly-refracting crystal). With this change comes a diminution of the inequality in elasticities, which, **at a** certain temperature, may even altogether disappear, as we shall hereafter demonstrate by a beautiful experiment, **due also to** Professor Mitscherlich (§ 171). The various faces **of** such crystals also show very different powers of cohesion; and even different resistances **to the** disintegrating action of chemical re-agents. Finally, it has been shown directly by Savart, who strewed fine dust upon plates of crystals cut in various directions, and then excited sonorous vibrations in the plates, that there are such differences in actual elasticity as we should expect.

If now we take a rhomb of Iceland spar and reduce one or the other of its sides till all the edges are of equal length —the true form of the crystal—we find it resembles quartz in being symmetrical around the one axis A A (Fig. 143), and no other. It is as if we took the skeleton outline of a cube made with wires, jointed at each corner of the cube, and laying one corner on the table, pressed down the opposite

---

[1] This experiment is due to Senarmont.

EFFECTS OF UNEQUAL ELASTICITIES.

one. In the longer rhomb, also depicted, the direction A A *parallel* to the other is still the true crystallographic axis round which the crystalline molecules are symmetrically built.[1] If we cut a plate with artificial faces perpendicular to this axis, and melt wax from a heated centre, as with the quartz, the melted area is circular.

Take now a ray passing along this axis. The **vibrations** being perpendicular to **the ray are** therefore perpendicular to the **axis, and in all these perpendiculars the** elasticities

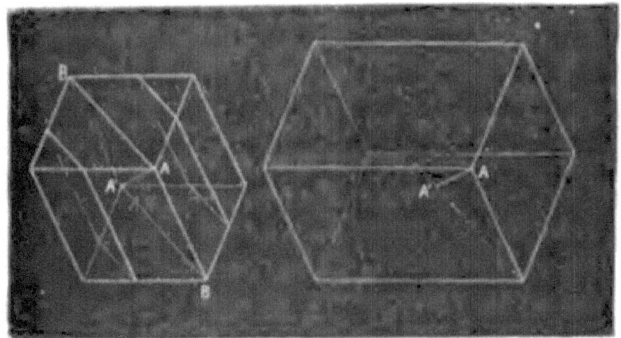

FIG. 143.—Axis of Iceland Spar.

are equal. There ought, therefore, to be no double refraction. If a ray is thus transmitted, we find it is so; there *is* no double refraction; and **the axis is** therefore called the *optic axis* of the crystal.

But now let the ray pass **through** the crystal at right angles **with** this axis. The elasticities being equal all round the axis, that **in the** direction of the axis itself **must** be either

[1] The axis **is a** mere *direction*, which at any point in the crystal, if it were split so **that** there was an obtuse solid angle at that point, would be an axis to that angle.

greatest or least—in the calcite it is greatest. The axis itself, in such a case, therefore, is a plane of vibration at right angles to the ray, and therefore such as luminous vibrations require, in which the ether vibrates most freely; and at right angles to this is another plane in which it is most retarded. According to the simplest mechanical principles, all the azimuths of vibration in the ray must be resolved into these two, and the ray be thus divided into two differently refracted, and oppositely polarised.

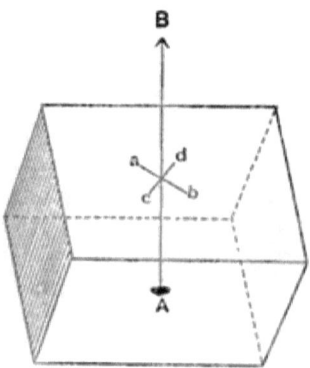

FIG. 144.—Direction of Vibrations in the Spar.

Take next an intermediate position; for instance, lay the rhomb on a flat table over a black spot; the ray sent perpendicularly in the direction A B (Fig. 144) to an eye over the spot is neither parallel with nor at right angles to the axis. But its vibrations, being necessarily at right angles to itself, may be represented by lines drawn on a piece of card laid on the top horizontal face of the calcite. Draw two such lines at right angles to each other to represent planes of polarised vibration, *a b*, *c d*. It will be found the card can easily be turned round so that there is a position in which *a b perpendicular to the axis*, the direction of least elasticity ; and the other line, *c d*, is that of greatest elasticity, and must, in this case, lie in the *same plane* as the axis. Into these **two** directions, therefore, will the vibrations be resolved ; and the two images will always be in the line *c d*, which is in the same plane as the optic axis.

122. **Principal Planes or Sections.**—The plane thus passing through *c d* and the optic axis, and all planes or sections parallel to **it, are** called *principal planes* in the **crystal.** Thus in Fig. 143, not only is the plane A B, A B a principal plane, but the other parallel planes drawn are principal planes. As therefore one of the two planes of polarisation, in a ray incident in **a** principal plane, coincides with that plane, and the other is perpendicular to the axis, the elasticities on both sides **of** the plane of incidence are equal, and both refracted rays remain in that plane, though differently refracted. It will be soon found on trial that all planes perpendicular to a natural face of the spar, and including the optic axis, are principal planes.

But if the **ray in** passing through the spar is oblique to both the optic axis and the faces of the crystal at which it enters and emerges, the case is different. Cut a small cross **of** card and stick it by a pin on the end of a thin rod, like a child's windmill, to represent—the rod the ray direction, and the arms of the cross two rectangular polarised planes. Take another rod and adjust it in a position to represent the optic axis. The latter being merely a direction, the *actual* axis any given ray is concerned with is that which intersects the ray ; move the optic axis rod (parallel to itself) therefore, till it intersects the rod representing the ray. It will be now found that **the cross** can always be turned round so that one arm is at right angles to the axis : this therefore, being the *least elasticity possible* in the calcite, represents one of the polarised planes of vibration, and being at right angles to the axis or plane of greatest elasticity, the forces on each side of that plane are equal, and this ray will follow the ordinary law of refraction. But the other plane of vibration is not now in the plane of the axis, or greatest elasticity, but **oblique to it.** The elasticities are therefore unequal on different sides, and this ray is deflected to one side or the other, the refracted ray not being in this case in the plane of incidence.

123. **Indices of Refraction.**—Lastly, it can easily be seen that in calcite the index of refraction for the polarised ray whose vibrations are in every case at right angles to the axis must be constant, and be **the** highest; **whereas** the index of the "extraordinary" ray must vary from that same index **down to** some lowest figure due to the greater ease of vibration in the direction of the axis itself.[1] In calcite the two indices are 1·654 **for** the ordinary ray, and 1·483 for the lowest value of the extraordinary ray. But in other crystals (as in quartz, for instance) the elasticity may be *less* **in the direction of the** axis than in directions at right **angles to it.** In that case the general phenomena **will be** the same, but the ordinary ray will have the *lesser* constant index of refraction.

124. **Wave-Shells in Uni-axial Crystals.**—We have thus accounted for the phenomena in crystals which have one axis of no double refraction. In both the cases (of which calcite is called a negative crystal, because the extraordinary ray is less refracted than the other, or the axial elasticity is greatest; while the other class are termed positive crystals) the *ordinary* ray proceeding from any point in the crystal would reach the same distance in any direction in the same

---

[1] It must never be forgotten that the direction of the *ray* is at right angles to both sets of *vibrations* we are considering. In calcite, therefore, the ray itself travels *slowest* along the axis.

time, and the wave surface may be represented by a spherical shell. But in the calcite the "extraordinary" wave-shell, similarly formed by the terminals of rays sent in all directions from the same point, will be a spheroid *flattened* in the direction of the optic axis; with its flattened surfaces just coinciding in the centres with the opposite axial **points of** the sphere, and its extended surfaces outside the sphere. **In** a positive crystal, on **the** contrary, the "extraordinary" wave-shell will be a *prolate* **spheroid**, of which the ends are coincident with the two axial points in the sphere, while the compressed surfaces are inside the sphere. The two may be

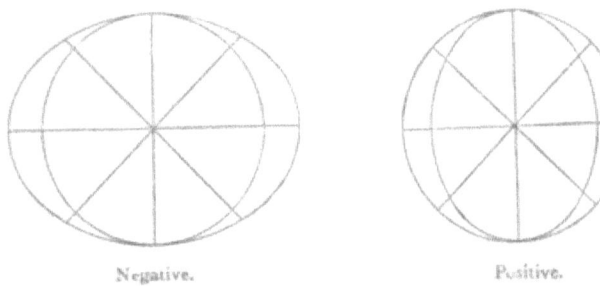

Fig. 145.—Positive and Negative Crystals.

represented in section as in Fig. 145, where the radii as they cut the two curves represent the respective velocity of the two rays;[1] and the optic axis **is** the perpendicular to the common tangent of the two curves.

As a rule, the spheroid coincides at two opposite points

---

[1] Of course the longer **and** shorter radii only represent the two velocities in the same precise direction. Any actual ray not passing along the axis itself must be divided into two separated by an angle, and these two separated radii will represent the velocities of the two halves of the ray. The precise geometrical construction for any given ray, as given by Huyghens, will be found in any of the text-books.

with the spherical wave-shell; or the two "indices" correspond in one direction of the ray. But in quartz and a few other crystals, as we shall hereafter find, there is a peculiar kind of double refraction in the direction of the optic axis itself, so that the spheroid does not reach to the spherical wave-shell, but is altogether contained within it. In this case also, however, the optic axis is a common perpendicular to *parallel* planes which are opposite tangent-planes to the spherical and spheroidal wave-shells.

125. **Bi-axial Crystals.**—Sir David Brewster discovered that there were many other crystals in which neither ray obeyed the law of ordinary refraction; and further experiment showed that such crystals possessed *two* axes of no double refraction. The phenomena of these crystals are explained on similar principles, but must be studied more in detail later on. In them also the optic axes are the perpendiculars to planes, which are common tangent-planes to the two wave-shells.

126. **Phenomena of the Tourmalines.**—We can now readily understand the appearances presented by the tourmalines and other apparatus. Considering the original beam of common light to consist either of vibrations in all azimuths, as at A, Fig. 137, or only of two at right angles with each other, as at B in the same figure,[1] which may, however, be at any angle with the horizon; in either case polarisation consists in the obtaining separately of vibrations in *one* definite path only; and "plane" polarisation (for we

---

[1] Either hypothesis will account for most of the phenomena, and with the two double-image prisms in one position, we did certainly compound one beam (F, Fig. 132) out of two rectangular plane-polarised beams, which cannot be distinguished by any test yet known from common light. Nevertheless the theory that common light contains all azimuths is far the most probable, and best meets certain important considerations. A brief summary of the matter is given in the Appendix to this chapter.

shall find other kinds) means that such path is in a plane. The phenomena then are very easily explained. Taking the tourmalines as an example, the original ray A (considered for convenience as compounded of two planes, or polarised rays, together) meets the first tourmaline, B (Fig. 146), so placed that the optic axis is perpendicular. We have already learnt that the ray must be divided into planes of vibration, one perpendicular and the other horizontal: but the horizontal vibrations are all absorbed within the crystal, which from some arrangement of its molecules not understood, is (when of a certain thickness) absolutely opaque

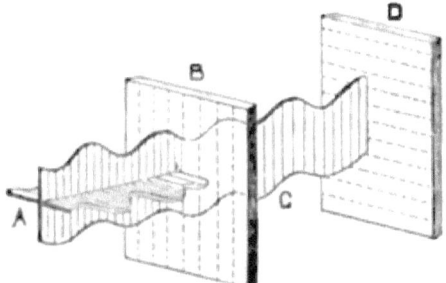

FIG. 146.—Action of Tourmalines.

to them. Therefore only perpendicular vibrations can get through. The ray which emerges, C, is therefore polarised, and if it meets a second tourmaline similarly placed, it will again get through. But supposing the second be placed as at D, all light must be stopped, as we have seen it is, though by two nearly transparent crystals!

The phenomena of the double-image prisms are explained in precisely the same way, except that in this case both rays get through the first crystal. When the second prism is at right angles with the first, the ray that got through the first would be quenched by the *same* plane of vibration in the second; but the plane at right angles with that allows it free passage. The phenomena of reflection and refraction do not need any further analysis; and it will only be necessary, before proceeding with our experiments, to describe briefly the most convenient polarising apparatus.

APPENDIX TO CHAPTER X.

The Vibrations of Common Light.

WE have only been able to account for the phenomena of polarised light on the supposition that the particles of ether moved in perfectly definite orbits; and every experiment in this branch of physical optics, from first to last, confirms that hypothesis, which we shall see has also enabled most marvellous predictions to be made, afterwards verified by experiment. But when we proceed to ask, What are the nature or orbits of the vibrations in a ray of common or unpolarised light, we have propounded a question the answer to which is by no means easy. All that can be done here is to state briefly the chief results of the investigations various physicists have made into this interesting subject,[1] and finally to suggest what appears the most probable method of reconciling the chief difficulties.

If common light passes through a doubly-refracting film it is divided into two plane-polarised beams; and if these are not much separated, or supposing they are, if

---

[1] It is right to state that great part of the following paragraphs are mainly a condensation, recast, of the admirable summary of the subject given in the last edition of Müller-Pouillet's *Lehrbuch der Physik*. A translation, or version in any sense, is not pretended; and the conclusion is not from that work.

they are again united, the two together behave as common light (§ 126). This may be and has been accounted for on the theory that common light itself consists of two plane-polarised beams, the vibrations of which are in rectilinear planes, so that a section of these planes across the ray would resemble A in Fig. 147. This seems to have been the belief of Sir David Brewster, and partly of Fresnel.

But this theory presents great mechanical difficulties. It is difficult to conceive that in any homogeneous transparent medium there should be any two given planes of vibration more than any others. Still more, either half of the ether particles vibrate in one plane, and half in the other, or else the same particle alternately vibrates in opposite planes.

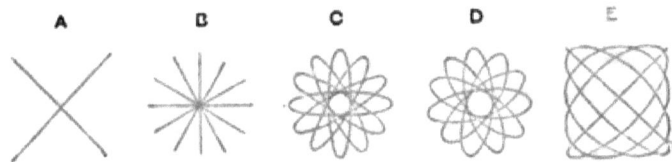

FIG. 147.—Theories concerning Common Light.

The last supposition cannot possibly be entertained; but even upon the other, *immediately contiguous* particles of ether must be vibrating in opposite planes. This is difficult to explain on any wave-theory, which supposes that each moving particle communicates *similar* motions to contiguous particles. And lastly, such a theory seems to suppose that there is really no unpolarised light at all, but solely bundles made up of oppositely polarised rays.

Fresnel therefore adopted the theory that the vibrations of the ether particles in common light took place in *all* azimuths, but that these azimuths were assumed *in succession*. On this hypothesis, taking for clearness only a few azimuths, a section of the ray may be supposed to be like B, Fig. 147.

In adopting or describing this theory it is often added,[1] that with such inconceivably rapid vibrations as those of light it is possible or probable that hundreds or even thousands of vibrations may take place in one plane, before they change into some other plane. But it is impossible, on reflection, to rest satisfied with any such supposition. If the vibrations thus remain constant in direction through any number, there is no reason why their direction should *then* change; and in fact whenever we have, as in a polarised ray of any kind, one which we *know* vibrates many times in succession in one orbit, we also know that such orbit of vibration remains stable.

Other considerations also require us to carry our reasoning further. If there be change, it must be either sudden (or by steps, as it were) or by insensible degrees. Obvious mechanical reasons are against the first supposition, and we are almost shut up, therefore, to the hypothesis of a gradual and continuous change in the path of vibration. But even if we suppose the elementary forms of such paths to be plane vibrations, such a gradual change must result in a curve; and when we remember that an *elliptical* orbit is of all forms of polarisation the most common in nature,[2] we are almost driven to take elliptical orbits into our purview. In fact Fresnel's own conception has been modified so as to consist of elliptical vibrations in various azimuths, resembling those of c, Fig. 147, rather than the plane orbits of B in the same figure. Such vibrations *gradually* changing azimuth would somewhat resemble the curve shown in D, Fig. 147. But the case is not only conceivable, but on every mechanical ground far the most probable, that any

---

[1] See *Polarisation of Light*, by W. Spottiswoode, P.R.S. (*Nature Series*), p. 6. See also Deschanel's *Natural Philosophy*, last page.

[2] Elliptical and circular polarisation are explained in a subsequent chapter.

original elliptical vibration should not merely change gradually in azimuth, but also in *character*, gradually passing not only into circles, but also into straight lines. Such a curve is roughly indicated in E, Fig. 147, except that in the case of light **we** must conceive the changes infinitely more gradual, and **so** arranged that each recurring **form** (*e.g.* each recurring straight line) should be in a somewhat different azimuth. In this way we can easily not only imagine, but with any compound pendulum apparatus readily trace, a curve which shall give in succession every form represented by Fig. 148, which it will be seen at a glance presents every known variety of polarised light.

FIG. 148 —Various kinds of Polarised Orbits.

Now such transition curves as these are actually known **to us**, being common to countless forms of acoustic apparatus. Two almost exactly unisonal tuning-forks with mirrors will produce them in the manner of M. Lissajous;[1] and they also occur in the *kaleidophone*, in the latter case especially being produced by the end of a rod in a state of *transverse vibration*.

Many ingenious experiments appear to favour this hypothesis. Common light may by proper means be divided into pairs of oppositely-polarised light of either of the three forms—plane, elliptical, or circular. Dove received a pencil of common light on a concave glass cone whose sides **met** the rays at the polarising angle. There thus met at the **centre**, rays plane-polarised in every azimuth: the result **gave** no trace of polarisation. He then passed

[1] See pp. 35-37.

common light through a swiftly-rotating calc-spar prism, thus producing a pencil polarised in a swiftly-rotated plane. Tested by a second prism, this rotated pencil showed no polarisation: tested, on the contrary, by the electric spark, it did; thus appearing to show that what we may call Dove's "artificial" unpolarised pencil differs from a natural one, chiefly in the greater *slowness* of the changes in its vibrations. Dove further caused a quarter-wave plate to rotate with the calcite polariser, thus producing circular and elliptic polarisation in all azimuths; the sensible result was still unpolarised light. Finally the quarter-wave plate was fixed while the polariser revolved, thus producing in succession plane-polarised light in two opposite planes, and also elliptically- and circularly-polarised light of opposite kinds. The result was still, sensibly, unpolarised light; but if the rotation was arranged so that the velocity in each revolution acquired two *maxima* and two *minima*, the sensible result appeared partially polarised. It is also well known that if a beam of plane or other polarised light be dispersed, or broken up by passage through many substances, such as a film of stearine,[1] it loses every trace of its previous polarisation.

All these considerations point to the theory which derives common light from vibration in some transitional figure of the nature shown in E, Fig. 147; but Lippich in 1863 started the objection that such curves, in acoustics, arose from combining rectilinear oscillations of very slightly *different periods*. Now in light-vibrations, such different periods would represent *different colours;* and accordingly he suggested that according to this theory no *absolutely* homogeneous or one-coloured rays of common light could exist. Lippich himself adopted the hypothesis that it was so, and

[1] Sir David Brewster gives a list of such depolari-ing substances in the *Philosophical Transactions*.

that all unpolarised light must and did contain periods sufficiently different to produce the compounded curves. This hypothesis is possible, and there appears no probability of any experimental means being devised sufficiently delicate to determine the point absolutely. Nevertheless it does not seem to be by any means necessary.

Briefly, it does not appear to me to have ever been really demonstrated that transitional curves involve as their cause two different *periods* of rectilinear vibration, in the particular case where the vibrating body is under the physical conditions represented by a point in a stretched cord. On the contrary, cords prepared with the most scrupulous care to avoid any sensible inequalities of tension or elasticity on any side, give phenomena of the same character; and countless experiments tend strongly to show that the character of the curve described by such points, depends rather on the particular *application of forces* to the point in the cord. It is quite possible to produce vibrations in a cord which remain nearly in one plane for a considerable period, and, by a different application of force, to produce most complicated curves. By threading silvered beads upon the strings, it has been shown that different performers cause the same string of the same violin to vibrate in different curves. We even know that, supposing a stretched cord or the end of a rod to be deflected radially by a given force, if a moment later any other force be applied tangentially to the former the result *must* be a curve of some kind. The character of these curves is also affected by surrounding conditions; for it has been found by experiment that the string of an old and fine-toned violin has a very sensible tendency to vibrate in more simple or "closed" curves, than that of a rougher, inferior instrument, thereby accounting for the purer tone.

Now there are some reasons to believe that the constitution of the ether may be such, that at any given point of displacement the forces acting upon it do resemble those acting upon a stretched **cord**. Moreover such vibrations afford the easiest **and best** conception of the possibility of the infinite number **of** periods **and** wave-lengths in **a ray of white light.** A stretched cord not only vibrates as a whole, but in **all** the segments or harmonics of which it **is** capable: and though these are limited in number in our limited cords, when we come to conceive a cord of infinite elasticity, and of such infinite length as compared with any wave-length of light as we must conceive in the case of waves in the ether, we at once account for waves **of** all periods being simultaneously propagated along the same cord, as the hypothesis requires. Further, even **with a mechanical cord of** the greatest possible length, wave propagation **(such as the transmission of a "shiver"** caused by striking near one end) **is one of** the swiftest we know which is capable **of** being traced by mechanical means. It is almost certain that, with molecular motions of sufficient rapidity to cause luminous rays, any given particle of ether may be, before it has ceased to vibrate under the influence of any given motion, influenced by another motion in quite a different direction. These motions may be complicated especially by displacements of the ether *in the direction of the ray itself.* These may act upon the transverse vibrations, though not luminous themselves, just as transverse vibrations in **a** cord are caused by a **pull at the** end.[1] There appears no difficulty, therefore, in conceiving **that transverse** vibrations may be perfectly synchronous, or produce **absolutely** homo**geneous light, and yet** assume transitional forms.

---

[1] Professor Stokes has shown that although longitudinal displacements o the ether particles cannot give rise **to** luminous phenomena themselves, it is to some extent necessary **to** take them into consideration in regard to certain results.

THEORY OF COMMON LIGHT.

In any case it appears far the most probable that the vibrations of common light, whether thus accounted for, or upon Lippich's hypothesis, do take place in such transitional orbits as are represented (in skeleton only) by E, Fig. 147, and that the forces acting upon the displaced ether-particles in a ray of light resemble those acting upon a stretched cord, unhampered however by its limited ends and other mechanical imperfections. The capabilities of this last supposition long ago struck Sir John Herschel; and it appears to me to cover more ground and to present fewer difficulties than any other. I have sometimes thought that it lends itself better than most other hypotheses to Clerk-Maxwell's, or, in fact, any other at present conceivable electro-magnetic theory of light. And it is very remarkable, to say the least, that if by means of smooth glass guides we compel any part of a cord to vibrate in a rectilinear or other fixed orbit, all other harmonic vibrations take place in a similar orbit; offering thus a striking analogy to the stable effects of polarising a ray of light.

CHAPTER XI.

POLARISING APPARATUS.

The Nicol Prism—Prazmowski's Modification—Large and Small Analysers—Nicol Prism Polariscope—Foucault's Prism—Care of Prisms—Glass Piles—The ordinary Lantern Polariscope—Simple Apparatus for Private Study—Norremberg's Doubler.

127. **The Nicol Prism.**—The tourmaline is an admirably simple polariser or analyser; and it gives a "field" of any angle, which makes it very convenient for "eye" work: it is however too small for large polarisers, and its colour is a serious drawback.[1] The polarisation by Iceland spar is also perfect; and as this spar is as clear as glass, if we can abolish one of the rays we shall get a colourless and perfectly polarised beam. This was effected by Nicol in the prism which bears his name. Reflecting that the two indices of refraction varied considerably (they are 1·48 and 1·65), and that Canada balsam, so much used for cementing lenses, was intermediate between these, he cut a crystal of spar thrice as long as its diameter through A B, Fig. 149, and cemented the

[1] The largest tourmaline known is in the possession of Dr. Spottiswoode, and is over two inches square. Strange to say, its colour and polarising properties are both remarkably good. It is difficult to get really good plates, tolerably free from colour, more than ¾ inch square, and such a plate is worth several guineas.

THE NICOL PRISM.

sections by a layer of balsam. Then a ray C D on entering the spar is doubly refracted, one half more so than the other. The *most* refracted half finds in the layer of balsam a less refracting medium; and obviously, if the angle is oblique enough **for** the respective indices, must be "totally reflected" to one side, and is thus got rid of. **The other** ray finds in the balsam a denser medium, and therefore passes through.

Nicol prisms are sometimes made with the end faces cut square to the axis. These need more spar, but we avoid part of the reflection from the inclined faces. Prazmowski further improved them by making the joint with linseed oil. There is difficulty in baking this dry, about which there seems some secret: but as the **"index" of** the oil is nearer

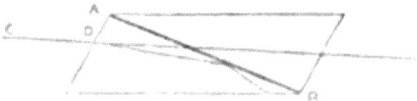

FIG. 149.—Nicol Prism.

than balsam to the lower one of the spar, an addition to the angular field is gained.

The "Nicol prism" is the most perfect piece of apparatus we possess, owing to the *absolute* polarisation of all colours alike, and its freedom from colour. One about 2 inches long will cost from 30s. to 50s., according to where it is purchased, and should be fitted in a tube which fits and rotates on the nozzle of the objective, (N, Fig. 1) to serve as an analyser. A large Nicol also makes the most perfect "polariser"; and the cost, which is many pounds, is its only objection. Large spar is scarce, and therefore dear; and a prism of 2 inches aperture requires a rhomb more than 6 inches long. If such a prism can be afforded, it should be mounted in front of the lantern flange, so that the full **parallel** beam passes through it, **and** so as to be capable of rotation. With such a polariser, we can turn the lantern direct to the screen, which is a great advantage; as also **is** the power of rotating the full beam as to its plane of polarisation. But, on the other hand, a large Nicol for *analyser*, as used by our chief lecturers, I consider a mistake in almost every way.[1] We must manifestly focus on the **screen all the slides we employ,** by a lens which will converge and cross all the rays. Now referring to Fig. 150, it will be seen that unless the **lens be of** long focus, the small Nicol, properly adjusted at the crossing point, allows every ray to pass through it which **can** possibly get through the larger one, however large **it be; and thus a small** Nicol so

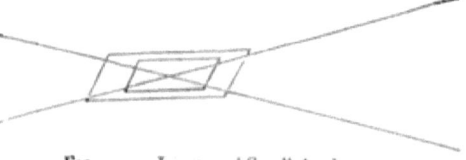

FIG. 150.—Large and Small Analysers.

adjusted actually saves absorption of light through a great length of spar. Moreover, **we can** adjust a small Nicol in this way precisely, and focus with precision by the rack and pinion of our objective; whereas with **a monster** Nicol in front, both the object and the focusing lens have, as usually mounted, to be coarsely **adjusted** by hand, and much light is scattered about **the room. Again,** to ensure getting most of the rays **through a long Nicol, a** lens of long focus has to

---

[1] A little qualification **is necessary,** because in some investigations it is desirable to carry a beam **of** *parallel* light through polariser, object, and analyser, to a prism or other apparatus; and for such experiments **a large** analyser has obvious advantages. **Such** experiments are however **few,** belong chiefly to scientific investigation, and are not suitable **for projection.**

be employed; which either necessitates a long screen distance, or gives a smaller image, and will hardly show any rings in crystals without additional and special apparatus. For these reasons, and after seeing experiments performed with large Nicols repeatedly, I adhere to an opinion based upon experience, that such a lantern-front as described in Fig. 1, with a small Nicol adjusted at the right point, **gives more light, a** better disc, greater facility in manipulation, and *superior effects* in almost every way.[1]

128. **Nicol Prism Polariscope.**—It **may be well to** describe a projecting polariscope (which is of course capable of private or eye-work also, by the light **of** a candle or that from a window,) made on this principle; and especially as I have never seen an instrument which, taken all **round,** was capable of such a variety of work with **so** little trouble and such rapid facility **in** manipulation, especially in the dark; a column of fluid, **or the** convergent arrangement for wide-angled crystals (see § 167) being added or withdrawn **in twenty** seconds. The adjustments have all **been most carefully** planned, several details of value being suggested **from their** own experience by Messrs. Darker, who constructed the whole with much care and skill from **a** Nicol of barely two inches

---

[1] It is scarcely necessary to state that nothing is further from the purpose of these pages than any depreciation of apparatus which is the admiration—almost the envy—of other than the happy possessors. The first pair of large Nicols made for Dr. Spottiswoode were of distinct scientific value, as being the precursors of this kind of apparatus, never before made on such a scale (they were about $2\frac{1}{2}$ inches clear aperture, and larger ones were afterwards made for the same gentleman, over $3\frac{1}{4}$ inches). And the larger a polarising Nicol can be secured or afforded, the better. But I certainly do desire to make clear to science-teachers and others, how very nearly all the experiments capable of projection by these magnificent instruments can be equalled on the screen at a far less prohibitory cost, not altogether beyond the means of some who might otherwise never dream of possessing such a class of apparatus.

aperture, but of great purity, made for me by Mr. Ahrens. The drawing is on a scale of three inches to the foot.

B (Figs. 151 and 152) is a base-board, barely twenty-four inches long, and six inches wide, in which mahogany slides, H (Fig. 151), are secured by screws, L (Fig. 152), which tighten the dovetailed or rebated slides. The screws bear upon slips of brass fixed to the slide-bases, H, as usual. On

FIG. 151.

these bases are screwed brass standards which carry separately the polarising Nicol, P N, the stage and focal power, S, F, and convergent lens system when required (Fig. 152).

The large Nicol, P N, is mounted in an inner tube, fitted in a larger tube—partly for appearance and partly to accommodate a rather larger Nicol should such ever be acquired. To save useless weight and space, its corners are cut off in a hexagonal form. The outer tube carries at the end next the lantern a flange, D ·(Figs. 151 and 152), divided into forty-five degrees, and at its middle four spokes, by which

FIG. 152.—NICOL PRISM PROJECTING POLARISCOPE.

D, divided flange (see fig. 151).
L, lens for parallelising convergent rays from lantern (a glass plate can be substituted in same fitting).
P N, polarising Nicol.
G, glass projecting cap.
S, spring slide-stage.
F, focusing power of two lenses, variable, the nozzle fitting all the apparatus.

C, converging lenses, with circular stage and springs to receive crystals.
Q, quarter-wave selenite, fitting in cork at pleasure, behind re-collecting lenses.
F F, variable focusing power for crystals.
A N, analysing Nicol, which also fits nozzle of F.
B, base-board.
L L L, fixing screws.

it can be rotated in collars screwed to the standards. Two of these spokes are bright, and two blackened, the latter also having a groove turned round them; so that the polarising planes can be known in an instant either by sight or in the dark. In the same end of the outer tube screws a fitting carrying the concave meniscus lens, L, or a plane glass to protect the spar, at pleasure. The object of the lens is, whenever the utmost illumination is required, to reconvert to parallelism the converging rays from an additional convex lens of similar focus, mounted in a frame as an ordinary slide, and inserted in the slide-stage of the lantern. Thus a four-inch beam can be brought down to a parallel beam of two inches; but usually a plane glass is employed, and G is another plane glass cap slipped over the smaller tube at the other end of the Nicol.

The slide-stage and power are shown drawn forward; but in ordinary use the end, S (Fig. 152,) is pushed up, so that the end of the Nicol, G, projects into S, close up to the slide-stage, and no light can escape. The stage and focal power are precisely the same as in Fig. 1; but two lengthening adapters, and an extra lens of six inches focus, either lengthen the focus an inch, or by using the back lens only, give also foci of five inches and six inches. These are very seldom required. A tube for fluids, eight inches in length, closed by glass plates with leather washers and screw fittings, is borne by forked standards on a mahogany base which drops into the base-board, but without dovetails, and can be inserted at G or withdrawn in an instant; or a flat-sided bottle or cell of fluid can be inserted with the same facility.

The analysing Nicol, A N, is mounted in cork, which is fitted into an *inner* tube sliding in an outer tube. The latter, for ordinary work, fits into the nozzle, F, as already described. By the inner sliding tube the analyser can be adjusted so as to come as nearly as possible at the actual crossing-point of the rays for various kinds of work. Much pains should be taken in selecting this analyser, many Nicols being cut so as to waste a great deal of angular field. Various prisms will differ by **a large** percentage, and one should be chosen which will let through (without showing the white limit to the field) as large an angular pencil as possible,—or, in other words, which **has** the widest (efficient) aperture for a given length. One and a half inches to two inches in length of side is the best average size. My analyser is one and a **half** inches in the side, and has an aperture the narrowest way of barely $\frac{5}{8}$ of an inch. With a focal power of about $3\frac{1}{2}$ inches, it just "clears," or shows without cut-off on the side, a circular slide of $1\frac{3}{8}$ inches diameter.

In Fig. 152, the analyser is shown **with** a convergent system of lenses for exhibiting the rings in wide-angled bi-axial crystals (see Chapter **XV.**). The first system of lenses, c, fits into the nozzle of **F,** and is provided with a circular stage barely three inches diameter, furnished with two ordinary micro-stage **springs**. This stage will carry any crystal, mounted anyhow, whether in wooden slides as hereafter described, on ordinary micro-glasses, or in the well-known square German cork mounts. The second system slides tightly into the end of a tube carried by a separate standard and mahogany base; and into the inner end of the short tube bearing it can be fitted at pleasure, a cork ring carrying the selenite quarter-wave plate, Q (see Chapter XIV.). The focal power, F F, for these lenses is carried by the inner end of a tube sliding into the other end of that carried by the standard; and being composed of two convex lenses with one concave, admits of **a** wide range of focus and consequent size of image. The other end of this inner sliding tube fits the analysing Nicol. The convergent lenses are in my instrument pairs of hemispheres. This is by no means the best optical arrangement; and in any other a wider angle would be secured by making the lens next behind each small hemisphere of greater diameter and somewhat larger curve. **I adopted** double hemispheres for compactness; and the system, though constructed **of** by no means dense glass, will bring both axes of crystals not exceeding 50° angle upon the screen. This point will be further dealt with in Chapter XV.

Besides these attachments, I have another tube, the small end of which fits into the nozzle of F, and carries a spring and slide-stage or slot, brought as near F as possible. Into the other end slides the tube carrying the two convex lenses of F F and the analyser. F F then forms an oxy-hydrogen microscopic power of $1\frac{1}{2}$ **inches focus,** or an inch power may be substituted. The whole light of the polariscope is condensed by the back lens **of F** upon the small slide-stage, protected as usual by alum-cells, either in **the** slide-stage s, the ordinary slide-stage of the lantern, or both, as required: **and** the polariscope thus arranged will **project** a large number of micro-slides—not all, but **a large class** which would not be shown **at** all by the ordinary **power.** This very simple **attachment** enormously increases the range of the instrument, bringing hundreds **of** cheap and easily accessible slides within its scope, particularly the more delicate sections **of rock. A section of porcupine quill, for instance, one-**fifth of an inch in diameter, can be projected at twelve feet **distance,** about 30 inches diameter. For higher powers there is hardly sufficient light.

The whole of these arrangements, **with Nicol,** double-image, **and thin** glass analysers, fluid tube, Huyghens' apparatus, and two ordinary crystal stages, **with a** small box for rotators and any special slides, packs in a case 30 × 10 × 7 inches; and the base-board will carry the whole arrangements, described in Chapter XVI., for projecting wide-angled biaxials through a tube of rotary fluid eight inches long (§ 184).

129. The Foucault Prism.—Foucault carried the idea of Nicol further; and by employing a film of *air* instead of Canada balsam between the two halves of the prism, made about one-third of the spar suffice which Nicol's construction requires. Fig. 153 shows the Foucault prism. The ray, A B, is doubly refracted as before, and one of the rays totally reflected; but air having so much lower a refracting index, the "cut" may be much less oblique to the ray. The aim is, as in Nicol's construction, to let one ray pass, while the more refracted one meets the air at an angle of total reflection. Owing to the less obliquity, however, there is less "angular field" in this prism; that is, a beam must be nearer parallelism

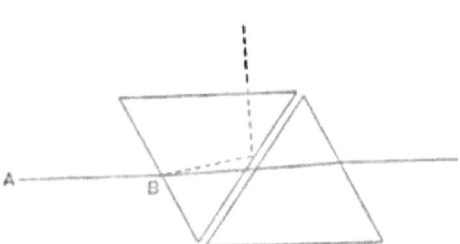

FIG. 153.—Foucault's Prism.

for all the marginal rays to get through—practically the convergence or divergence **must not** exceed about 8°. Also, the two surfaces of the film of air cause a perceptible loss of light by two partial reflections, even in the ray which does get through. This is not very great, however; and a large Foucault prism is only about one-fourth the cost of a Nicol of the same aperture.[1] I have never tested

---

[1] Besides requiring only a third of the spar, and much spar being available which is useless for a large Nicol, there is no troublesome balsam-joint to make, the cut faces only needing to be polished. Mr. C. D. Ahrens, who has made all the gigantic Nicols for Dr. Spottiswoode and others, has informed me that the joint of such a Nicol may

the matter, but am inclined to think the loss of light as compared with a Nicol would be about 10 per cent.

130. **Care of Prisms.**—Calcite is very soft, and all prisms made from it, of any kind, need special care. Nicols are often mounted with glass caps, which protect them while enabling them to be used; but if light is rather deficient these may have to be removed; caps should also be taken off for a few minutes in cool weather, to allow the dew to evaporate, which at first condenses when the light is turned on. If any dust or dulness has to be removed, the surface should be cleansed *solely* with a camel-hair brush of the finest quality. Small ones are best cleaned with a swan quill cut rather short; large ones require a round-nosed "sky-brush" of double that diameter. The application of chamois leather, as to lenses, would soon cover the spar with scratches. Another rather important point with *large* Nicols, is always to put them away with the balsam-joint standing perpendicularly. If one half rests upon the other, the weight of the top half may gradually produce "thin film" colours in the layer of balsam. Mr. Ahrens tells me several fine Nicols have suffered for want of this precaution, and have had to be taken apart and re-joined.

131. **Polarising Glass Piles.**—The next best polariser is a bundle of glass plates. A brass or tin elbow must be made, as Fig. 154, one end, N, fitting on the flange-nozzle of the lantern, the other end as short as possible, ending in a screw-collar, B, exactly the same size and thread as B in Fig. 1, so that the objective and optical stage will screw into it. At the corner of the elbow is fitted the pile of oval glass plates, G, made of the thinnest and most colourless glass procurable. Ten or twelve are sufficient, but the back

require thirty hours of unremitting attention over a fire, as it is impossible to leave the task when once commenced, till the balsam is baked thoroughly hard.

one must be blacked on the back, and the whole so arranged that the light from the condensers impinges upon and leaves the glass at an angle of about 56° from the normal. Such an elbow and bundle, with the objective screwed in, and a Nicol fitted on the nozzle, is the ordinary "lantern polariscope." It gives a large field of polarised light, and for nearly all purposes the reflector is almost as good as a large Nicol; but it has the two disadvantages, that the lantern has to be turned off sideways, on account of the elbow angle, and that the original plane of polarisation cannot be rotated, which is desirable for some experiments. Such experiments are, however, very few. To meet this objection,

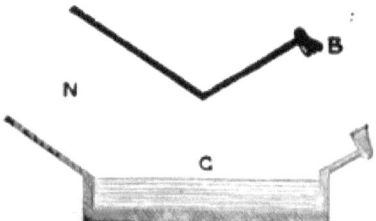

FIG. 154.—Elbow of Polariscope.

some have employed a glass pile, working by the transmitted light, which can be mounted and rotated just like a Nicol. This meets the objection just stated, but has another—viz., that it is difficult to get *complete* polarisation in this way. If a refracting bundle is employed, not less than eighteen plates should be used, and the angle should be somewhat *greater* than the polarising angle, by which nearly all unpolarised light may be reflected and so got rid of. The greatest fault of such a pile so used, however, is that it usually gives a perceptible green colour, owing to the thickness of glass. The "ten or twelve plates" often mentioned, do not polarise the whole beam by a great deal. A reflecting pile yields about 20 per cent. less light than a Nicol of equal field.

It is well to have an additional glass analyser, (Fig. 155,) formed of eighteen or twenty pieces of microscopic thin glass, G, placed at the proper angle in a tube which fits at N on the nozzle of the objective. An aperture, R, in the side of the tube, allows the reflected ray also to be used. For reasons already given, such an arrangement is not equal to a Nicol; but it gives a larger field, is very instructive and interesting in throwing complementary images of *all* phenomena on screen and ceiling, and will do *thoroughly satisfactory work* for all to whom 30s. or 50s. for a Nicol prism

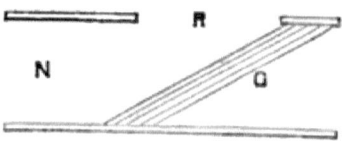

FIG. 155.—Thin Glass Analyser.

may be an object. In fact, a glass reflector and thin glass analyser are within the reach of any one who can work in brass.[1]

132. Table Apparatus.—For merely private study very cheap and simple apparatus will suffice. Make a shallow pasteboard or wooden tray, A (Fig. 156), 1 inch deep, say 7 inches by 4 inches. Drop into it thin glass plates to within ¼ inch of the top, cleaning them well, and blacking the back of the bottom one. Cut two wooden or pasteboard side-pieces, E, united across the top by another piece, D, and fill in the ends by a piece of *ground*-glass, B, 4 in. square, and another, C, of *clear* glass the same size. The light will fall on the ground-glass as shown by the arrow, if it is turned towards a

---

[1] My own *first* glass analyser was fitted up by myself, and used, in a pasteboard tube.

lamp or the window; will be nicely **softened** and scattered, and polarised by the pile in the bottom; and the objects can be laid on the clear glass at C to be examined by the small pile of microscopic glass, P, contained in a round or square cardboard tube.[1] The lantern polariscope already described **also makes a** capital "table" polariscope if a round disc **of** ground-glass, to soften the light, be fitted **into** the end, N **(Fig.** 154), which **goes** on the flange of the lantern. It will be found that the lenses pleasantly focus the objects in the slide-stage, which are examined **by** looking in through the analyser on the nozzle.

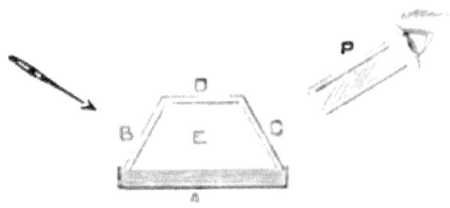

FIG. 156.—Simple Table **Polariscope.**

In fact, a small Nicol, or tourmaline, or pile of thin micro-glass fixed slantwise in a tube, **as** analyser, and the light reflected from any glass plate, or the top of a mahogany table, as polariser, will suffice for many experiments.

133. **Norremberg's " Doubler."**—There **is one form** of polarising apparatus which is **so** useful if the student **does** any personal work with thin films, as to need special mention. As designed by Norremberg, it is as in Fig. 157, where the ray of light, shown by the arrow, is reflected at the polarising angle from the *single* plate of glass, F, perpendicularly to the horizontal piece of looking-glass, H. It is thence reflected perpendicularly upwards, passing this time

[1] A cheap polariscope, constructed on this principle, is sold for 6s. **by** Messrs. Griffin and Sons, Garrick Lane.

through the polarising plate to be examined by the Nicol, or other analyser, at N. It will be obvious that if the object be laid on the stage at E, the phenomena are as usual. But if it be placed between the polariser and the looking-glass, at D, the polarised ray has to pass *twice*

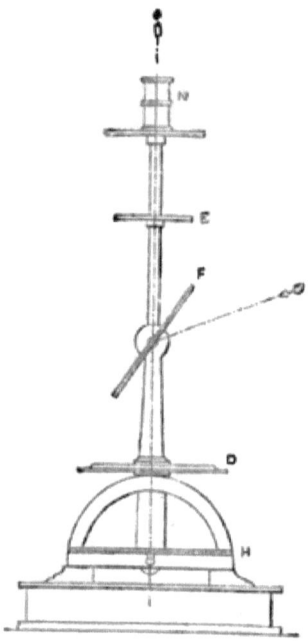

FIG. 157.—Norremberg's Doubler.

through the film or object to be examined, which is equivalent to doubling the film in thickness. The use of this "doubling" in ascertaining the thickness of thin films, will appear in the next chapter; and there are other peculiar uses of this form of apparatus. A very simple construction will answer all real purposes, however. Knock out the opposite sides F and B of an oblong box—such as a cigar-box —and on the remaining sides fix guides for the polarising plate, P, at the proper angle with the perpendicular. On one end, R, lay a piece of good looking-glass the size of the end; and in the other end cut a hole in which the Nicol, N, or other analyser, can be rotated (Fig. 158). The object can

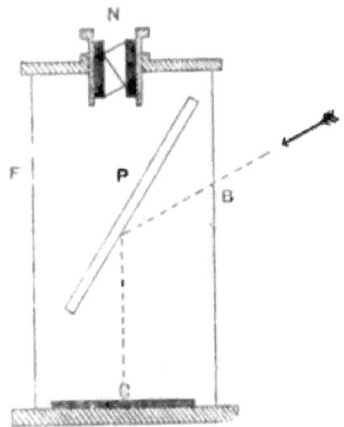

FIG. 158.—Simple Doubler.

be held between P and R, or even laid on the looking-glass itself in many cases. For most purposes of the "doubler" no focusing lenses will be required. I have even laid a piece of looking-glass on the table, and arranged over it a plate of clear glass at the proper angle by means of the Bunsen universal holder (Fig. 12), holding the Nicol in my hand.

NOTE ON ARTIFICIAL NICOL PRISMS.

The second half of a Nicol having no doubly-refracting effect, I have sometimes thought that large spar might be made to do double duty by making the second half of *glass* of the proper density. MM. Jamin and Soleil have employed an oblong cell with glass ends filled with bi-sulphide of carbon, across which was slanted a mere film of calcite. The bisulphide might be brought to exactly the higher index by benzol, and it will be seen that with this apparatus the action of the Nicol is precisely reversed, the total reflection taking place in the calcite. But though theoretically perfect, this apparatus is dangerous, owing to the heat of the lantern.

CHAPTER XII.

CHROMATIC PHENOMENA OF PLANE-POLARISED LIGHT.—
LIGHT AS AN ANALYSER OF MOLECULAR CONDITION.

Resolution of Vibrations—Interference Colours—Why Opposite Positions of the Analyser give Complementary Colours—Coloured Designs in Mica and Selenite—Demonstrations of Interference—Crystallisations—Organic Films—Effects of Strain or Tension—Effects of Heat—and of Sonorous Vibration.

134. **Resolution of Vibrations.**—We could have formed no conclusion as to the precise orbits of the molecules of ether in our polarised waves, apart from the phenomena; but if we have rightly interpreted these, then any one acquainted with elementary mechanics and the "resolution of forces" will see that we can test such a theory by experiment. Dealing with motions whose direction we are supposed to know, if we are correct we can "resolve" those motions. Taking a perpendicular plane of vibration, for instance, and supposing our ether-atoms move solely in the plane orbit A B, B A (Fig. 159), if we interpose a plate of crystal which only permits of vibrations in the directions B C, B D, the perpendicular plane must be resolved into *two* planes at an angle of 45° with the original plane.

It seems evident already that we *have* done this, from the duplication of our images when the two double-image prisms were at an angle of 45° with one another, and from the transmission of half the light through our two tourmalines at the same angle. But if we are right, the two oblique planes must be also capable of resolution in their turn, by the analyser, into perpendicular and horizontal planes; and when our polariser and analyser are crossed and the "field," therefore, quite dark, if we interpose between them a tourmaline at 45° we ought to *restore* the light. Cross therefore the Nicol or other analyser till the screen is dark, and insert the rotating tourmaline. Sure enough, as we rotate it, though the tourmaline is really a brown tint

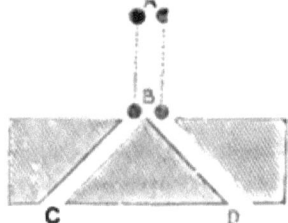

FIG. 153.—Effect of a Crystalline Film.

mounted on a clear glass, it appears as a *light* image on the dark field.

We have, however, learnt that the tourmaline stops one of the rays, and we wish to see beyond doubt if the original plane really is resolved into *two* planes. We therefore want a slice of some crystal which allows both halves of the doubly-refracted ray to pass, and either selenite or mica is convenient, as splitting easily into thin plates, which contain both planes of vibration at right angles to a ray transmitted through them. A crucial test suggests itself. In F, Fig. 132, we re-compounded *one* beam out of the two beams from our first doubly refracting prism. We see, on reflection, that if each of the first two beams is really again split by a properly adjusted film into **two**, vibrating in planes at 45° angle with them, that is no longer possible, and there must remain at least *three* images. We find, then, by experiment with polariser and analyser crossed, the position of the slice of selenite which still leaves the field dark, and which, of course, **gives us** its polarising planes. We then mount a circular disc of it in a position at 45° angle with that, in a **wooden** slider **an** inch wide, and 4 inches long, with a circular **hole in the centre** for the slice, protected between **two** discs of glass. With the two double-image prisms giving the single beam F, we introduce the slider into the slit s provided for it in our Huyghens' **apparatus, Fig. 131.** Our expectation is justified to the letter; for *three* images at once appear on the screen.

135. **Interference Colours of Polarised Light.**—But here we have another beautiful phenomenon. If the slice of mica or selenite be thin, these are *coloured* images, and each pair presents complementary colours; for if the images overlap anywhere, we get there white light. These coloured images are singularly beautiful as the front prism is rotated; and that is why the Huyghens' apparatus was recommended to be arranged as described.

We can readily understand this colour. In every bifurcation of the ray, it is doubly refracted because of unequal elasticity, and one ray is *more retarded* than the other. We can "see" this in a large piece of Iceland spar, for one of the images of a black spot seen through it appears nearer than the other. The two rays, while they vibrate in planes at right angles to each other, cannot of course interfere; but the analyser brings portions from each again into the same plane. Now in the rigid plane orbit of our original polarised beam we have that *identity of origin* we have already learnt (§ 96) is necessary for two rays to interfere; and in the absolute plane into which the rays separated by the selenite are again united by the analyser we have that *identity of path*, or nearly so, we also found to be necessary. We bring together **again**, then, into the same plane (that of the analyser) two originally identical rays, one of which, during separation, has got behind the other in passing through the **film by** a given distance, depending **on** the thickness **of** the film of crystal. But whilst in reflection from **a** "thin film," one **ray is** retarded by *twice* the thickness of the film; in this **case** one ray is retarded by the *difference* in velocity, whilst both traverse the **same** film. Of course a much thicker film is required in this latter case; and, of course also, the greater **the** difference in the two indices **of** refraction, the thinner the film must **be** to produce **a** given colour. Too great a thickness, of course, gives **no** colour; for the same reasons **too** thick a "thin film" gives none (§ 101).

136. Cause **of Complementary Colours.**—To explain the "complementary" **colours, we must** take into account the *direction* as well as the plane of vibration **of the** ether-atoms, **at** each moment of bifurcation or resolution into two planes at 45° angle with their path at that moment. Let us suppose the original plane-polarised ray A, Fig. 160, is at the phase when the atoms are moving downwards when it meets the selenite with its planes at 45°. The bifurcated **rays** must obviously travel in the directions B and C. Now B, when again resolved by the analyser, must take the directions D and E, and C of F and G. A double-image prism would transmit both, but by our Nicol analyser when crossed or parallel one plane or other is stopped. It is readily seen that when in the position which allows the two *perpendicular* vibrations to get through, these two (D F) are in the same phase of their orbits, and so coincide with or strengthen each other; but if the horizontal vibrations get through, E and G are in *contrary* phases, or destroy each other. However, therefore,

the two sets of waves come together in one position of the analyser, as regards any colour, that colour must meet in exactly *opposite* phases in the other position, at right angles to it. If the two waves are exactly destroyed in one, they are fully combined in the other; if half destroyed in one, they are half reinforced in the other, and so on. Therefore the totals must result in complementary colours.[1]

It is further evident from this, that in addition to any retardation or difference of phase due to the difference of velocity in the two rays while passing through the doubly-refracting film, when the polariser and analyser are *crossed* there is an additional difference of phase of *half a wave length*, since E and G, irrespective of any *thickness* of the

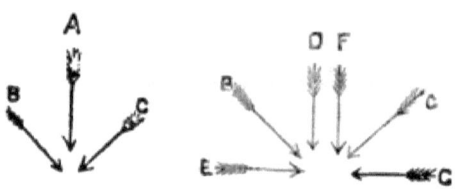

FIG. 160.—Resolution of Vibrations.

film, are in opposite phases. We shall see that this fact is of great assistance in accurately measuring films (§ 137).

It is also evident that if the analyser is turned round 45°, there can be no colour, because, of the two rectangularly polarised waves from the film, one passes through the analyser unchecked, and the other is stopped. Similarly, when the planes of the film coincide with and cross that of

---

[1] It will be seen in a subsequent chapter that the rectangular vibrations emerging from the film are chiefly compounded into circular and elliptical orbits. These are, however, again resolved by the analyser into rectangular planes; and the student will easiest grasp the subject at this stage by confining his attention to this simpler representation of it.

the polariser, there is no effect at all, the plane-polarised ray passing through unchanged, as if the film was not there.

137. **Coloured Designs.**—A film of varying thickness must of course give varying colours between the limits of colour. **A thin** slice split irregularly from selenite soon shows that this is **so ; and** thus if we have designs ground away of various thicknesses, **we** may form stars, butterflies, flowers, birds, &c., of their appropriate colours, which they owe **to** nothing but the interference of polarised light ! Such designs are prepared at all prices **from** 3*s.* 6*d.* to 3*l.* 3*s.* in selenite, and there is a strange fascination **about** them, changing as they do **to** complementary **colours at** every quarter-revolution **of** the analyser, and giving, **as** explained, no colour **at** all when that is at an angle of 45°. A **plate** ground concave gives of course "Newton's rings," **and** a plate ground slightly wedge-shaped must similarly give straight parallel coloured bands. Such wedges may be ground with water **on very fine** ground glass, and afterwards mounted in balsam between two glasses. Smaller reductions of thickness may **be ground in** selenite **with** the rounded end of a slate-pencil **and** some putty-powder, and polished with more putty-powder on a piece **of** wash-leather.

But the best crystalline film for students is *mica*, the same as used for gas-light covers, and quite a quantity of which **can** be bought for a few shillings. Choose a slab as clear and even as possible. This splits easily—almost too freely—into thin laminæ, and is easily cut through by a penknife. Such stars, of different colours, as A, and such concentric circles as B, Fig 161, are therefore easily made by cutting through a small thickness, and splitting off a film from each ray or circle. They can then be mounted in frames; and though the smokiness of the mica when made in this rough and thick way makes them rather inferior to selenite, the effect is still good. Again, mica is easily scraped away with a penknife to gradual thicknesses; and hence simple designs of tulips, fruit, &c., are pretty easily made, keeping the mica meantime on such a table polariscope as is shown in Fig. 156. Where scraped, the dry mica appears semi-opaque; but when mounted in balsam this quite disappears, and all is clear again.

A very much better way, however, is to first split the mica into very thin even films, as large as possible. They can be split so thin that the retardation is only one-eighth of a wavelength; and one-quarter is easily obtained. Then having found and marked on the sheet the polarising planes, various

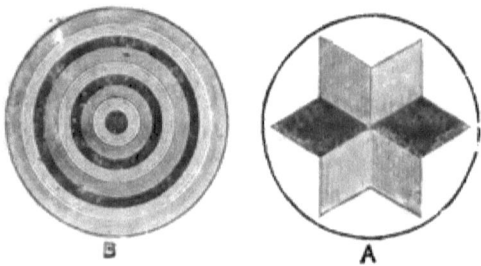

FIG. 161.—Mica Designs.

designs can be cut out with sharp scissors, to be *superposed* on one another, on the principle that every added film, as superposed with the polarising planes in proper position (§ 138) will alter the colour. In this way very attractive geometrical designs are easily made, which have the advantage over those described above of showing sharp and clean edges, and are much clearer owing to the smaller thickness of mica required. These thin films are best mounted in balsam and benzol,[1] only superposing four or five at a time, and letting them dry a little before adding more, or putting on the

[1] The "balsam and benzol" sold by opticians cannot always be depended upon to dry. The proper fluid is made by baking old balsam in a wide vessel, in a slow oven, till it is as dry and hard as

Pl. 4 INTERFERENCES OF POLARIZED LIGHT.

A Wedge of 24 thin Mica films.   B Ditto with plate of Mica crossed.
C Wedge of 8 Mica films.   D.D Two wedges crossed.
E Complementary Images and Spectra of Selenite films.
F¹ F² F³ Spectra of Selenite wedge, from thin to thick edge.
G Square of chilled glass   H Square of glass compressed in vice.

covering glass; otherwise the films slip about and spoil the preparation. I was indebted for the knowledge of this method to Mr. **C. J.** Fox, F.R.M.S.; and as an example of what may be accomplished by it, may mention a beautiful slide **which**, by means of twenty-four superposed films, each $\frac{1}{18}$ inch shorter than the one beneath it, gives the first three orders of Newton's colours, each divided *exactly* into successive $\frac{1}{8}$ wave-length retardations or divisions. **(Plate IV. A.)** Such accuracy in the thickness of the films must be attained by trial in Norremberg's "doubler." Obviously two plates of $\frac{1}{8}$ wave retardation, superposed the same way, must make a quarter-wave. The "doubler" makes this equal to a *half*-wave. And the crossed Nicol (§ 136) gives another half-wave. But the whole wave is known by almost perfect transmission, while half-wave retardation is known by almost perfect extinction, leaving only a dull "transition-tint" of reddish plum-colour between the first and second orders of colours. If therefore two films, on the mirror of the "doubler," with analyser parallel, give this tint, and it is known by the general "feel" of the films to be between the *first* red and the adjoining blue, they are known to be $\frac{1}{8}$ wave films.

138. **Demonstrations of Interference.**—That the colours really are produced by the greater retardation of one ray in the film, and subsequent interference, may be proved by several independent methods. First, we can easily stop one half of the bifurcated ray. Place in the stage any

---

pitch, and will chip into flakes. It is then to be dissolved in benzol (*puriss*). A drop or two is usually sufficient between each film, put on in the centre. When the preparation is finished, it must be *left for some days before baking*; a week is not too long. This is the only way to avoid bubbles. It can then be baked in **an** oven, or on the hob, till dry. The mounting makes the fine scratches, &c., invisible. Only a small weight or pressure must be employed, or the films will be displaced, and air may enter when the pressure is removed.

slide—say a concave Newton's ring slide. The rings being at their brightest, and with the analyser exactly crossed, introduce, in front of the selenite, the rotating tourmaline at an angle of 45°. **This stops by absorption** *one* **of the two** oblique rays **into** which the original polarised beam **is** divided; there are no longer two rays to be brought into interference, and accordingly, over the area covered by the tourmaline, the colours disappear.

Secondly, we may retard the other ray, and thus bring **the two again** into coincidence. If one ray be retarded in the selenite more than that vibrating at right angles to it, it **is** plain that, taking two selenites of equal thickness, if both are superposed the same way of the crystal, the colour must be that of a plate equal to the sum of both; **but that if** one be turned round 90°, the retardation of one ray **by the first must** be neutralised **by the second, and no colour at all produced.** If we place **two similar** films in the stage, **in** the two different separate positions, **we find** that it is so; and similarly, if the films be of different thickness, the colour will be in one position that of their **sum,** and **in** the other **of their** difference. A more striking demonstration is furnished **by rotating a** wedge over a similar one. The colours differ remarkably according to their positions; black (when the analyser is crossed) being necessarily produced wherever **exactly the** same thicknesses come into rectangular **positions,** and so causing in one position a black diagonal line. There**fore if a concave plate be rotated over a convex plate, the** phenomena **of the rings vary in a beautiful manner,** black circles appearing **in certain positions; or an even film** of the right thickness, rotated **over the concave, will give** the same beautiful **phenomena.** Again, if a film of even colour is rotated over a star, with the points in different positions, the colours of the points are affected very differently, and in what **appears (seeing the rotated film is** the same thickness all over) a most wonderful manner, till we understand the reason. The most instructive and pretty demonstration of these reversed or counteracted retardations is, however, furnished by two precisely similar "step" wedges built up on Mr. Fox's plan, out of films $\frac{1}{8}$ wave to $\frac{1}{8}$ wave in thickness.[1] The glass discs of the polariscope will nicely take a wedge $1\frac{1}{4}$ inches long by 1 inch wide, and this breadth is conveniently divided into eight steps $\frac{1}{8}$ inch wide. It is sufficient if each *corresponding* step in the two wedges is cut from the same film. One wedge being mounted in wood, and the other used in the rotating frame, when one **is** superposed on the other with the polarising planes in same positions and the thickest part of one over the thinnest in the other, we have an even colour. When the two thickest sides are superposed in the same positions, if the planes are parallel the tints grade with double the amount of difference in colour; if the planes are crossed there is no colour at all. When the wedges themselves are crossed, in one position there must be a central diagonal row of black squares, the other squares giving a beautiful chequer pattern of various colours. And when the rotating wedge is diagonal, there will be a pretty pattern of backgammon points. Two wedges built up of similar even films thus offer one of the most fascinating and instructive polariscope combinations. (See C, D, Plate IV.) Or should the student have sufficient skill and patience to build up a wedge of 24 $\frac{1}{8}$ wave films, if a single film be superposed on it with its principal plane crossed, and of a thickness equal to the middle stripe, it is plain that middle stripe must be black; and as on each side of it each stripe must present an equal *difference* or essential thickness,

---

[1] The first few $\frac{1}{8}$ wave thicknesses giving poor colours—only pale fawns and blue-greys—the broadest or foundation film of the wedge should be thicker, choosing the thinnest which gives a fairly pleasing tint.

the coloured stripes will be symmetrically arranged on each side of the black one.[1] (Plate IV. B.)

Lastly, we can prove the cutting out of certain colours by interference, by our never-failing method of spectrum analysis. Place a slit in the stage with a film which shows colours, and pass the light from the analysing Nicol through the bisulphide prism. There will be crossing the spectrum one or more dark interference bands—more with a thicker film; in fact as the film increases, more and more bands appear in various parts of the spectrum, just as in the light reflected from a film of mica (§ 101), and showing in the same way how with a certain thickness we fail to get colour. This is well shown by subjecting a wedge to spectrum analysis, a slit being placed in the stage, and the wedge, with its bands perpendicular, gradually advanced over it from the thin edge to the thick one. It will be seen that, as the thickness increases, a greater number of the interference bands cross the spectrum (Plate IV. F^1, F^2, F^3), as we should expect, accounting for the paler colours. It is also seen, as before, that in contrary positions of the analyser, the bands occur in complementary colours. The two complementary sets of bands are easily shown together by using a double-image prism as analyser, with the two images of the slit perpendicular. The slits will then overlap in the middle of the double image, giving there a white slit and complete spectrum; while the two complementary spectra will appear at the top and bottom. (Plate IV. E.) Lastly, if the slit be adjusted so as to cross the centre of a concave film giving Newton's rings, and the slice of light be analysed by the prism, we get again exactly the same interference bands shown in Plate III. B.

---

[1] It may be doubted if such accuracy is possible. It is what I have seen produced by Mr. Fox's own hands, and have repeated myself with ¼ wave films.

139. **Crystallisations.**—Another beautiful series of objects showing gorgeous colours in the same way are *crystallisations*, from thin films of various solutions flowing over glass discs, and then crystallised by evaporation. The only difficulty is to obtain the right thickness, as both too thick or too thin give little or no colour, though light and shade effects can always be had as the analyser is rotated. For this reason, what are called "superposition films" are very useful. These are thin plates of mica or selenite which give uniform colour (such are easily split from mica) mounted between two glasses. One which gives blue and yellow, and another red and green, should be provided; then a film of crystals, or anything else which shows no colour by itself, will give great variety when superposed on the film. There is a sameness about the colours produced by this plan, however, inferior to a slide which can show its own colours. A good arrangement for making crystallisations by evaporation is shown in Fig. 162. On a wire tripod is placed one of the glass candle-chimneys so common, 3 inches diameter, and on the top of this is laid a square metal plate, not quite covering the chimney, having in the centre an aperture nearly as large as the glass discs, with three little projections bent up $\frac{1}{16}$ inch from the inner edge to keep the glasses in place. A spirit-lamp underneath gives a steady and calculable heat, adjustable by raising or lowering it. A saturated solution is not always best; often one mixed with equal bulk of water does better, and sometimes a little alcohol added helps effect. Great interest will be found in preparing slides with weaker or stronger solutions, and less or more heat, which will often entirely alter the character of the crystallisation. Thus a solution of tartaric acid evaporated in the cold often crystallises in long straight lines, and in the sun in "stars;" whereas, when evaporated over the lamp, it gives irregular facets, very beautiful on the screen,

and larger as the heat is greater. Salts which give a pattern too small in this way, may often be made to give much larger crystals by dissolving gelatine or gum in the saturated solution, and evaporating in the cold. The following are good polarising salts :—Tartaric acid, and most tartrates ; citric acid, and most citrates ; oxalic acid, and most oxalates ; borax ; chlorates; many nitrates ; picric acid ; sulphates of copper and magnesia, and the two mixed ; most

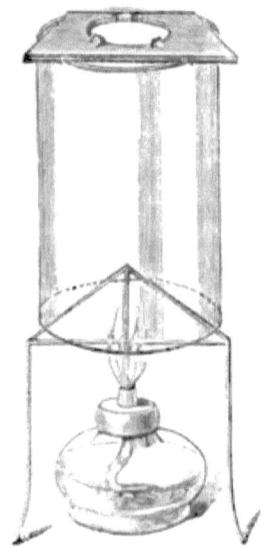

FIG. 162.—Apparatus for Crystallisation.

of the alkaloids and their salts; sugar—but in fact the list is interminable. Cubic crystals, being of equal symmetry and elasticity in all directions, do not polarise, unless in drying the film of crystals becomes subjected to tension, as it sometimes does (§§ 121, 141).

One of the prettiest crystals is salicine, which gives an enormous variety of effect, according to how it is treated.

Some salicines I have prepared **have** been particularly admired for their gorgeous colours and size of the crystals. They are attained by dissolving the substance in one part alcohol to four parts water, made rather hot, and saturated. Pour a *good* layer of this fluid on **the** glass, and evaporate quickly with rather a strong heat; the salt then *melts* in its own water of crystallisation, and beautiful crystals soon begin to form. The heat then needs humouring in a way only experience can give, else crystals already formed may be re-melted and the slide spoiled; but when all goes well the result is simply glorious, the whole slide being covered with circular crystals showing sectors of colour like miniature " Newton's discs," and each, when the analyser is crossed, exhibiting a black cross. (Frontispiece, **A.**) **The same film,** cooled and breathed on whilst the crystals form, **gives** quite different phenomena, and a thinner film different again. Almost **every** operator has some little secret **of** his **own**; and **one of my** correspondents sent me a **slide of the** small sort of singular **beauty,** prepared with gelatine. **The effects** range from discs $\frac{1}{16}$ of an inch to 1 inch in diameter, of *any* colour, or black and white, according to the thickness. Citrate of magnesia may be made to give very similar discs. Chlorate of potass crystallises in square tablets; nitre in long thin prisms which appear as a network of coloured threads. To get this latter effect the solution must be rather dilute, and after the disc is covered, all the solution that will go, jerked off; then left to evaporate in the cold. Most of the salts may be either mounted in balsam or simply covered with another dry glass and mounted with red putty in wooden frames, choosing the best positions before fixing. Deliquescent substances, like tartaric acid, are better in balsam.

Another class of beautiful crystallisations is formed by *melting* the substance between two glass plates. Make a pair of spring-wire forceps, like Fig. 163; then put some of the chemical between two clean *plate* glass discs, and hold in the forceps over a spirit-lamp: or over the chimney of an Argand gas-burner is not too hot for some. Care, of course, is needful not to crack the glasses. Most of the substances that crystallise well in this way are of the "organic" class, and some require a very thin film, which must be obtained by putting down the two hot glasses on a blotting-pad, and with thick padded gloves working close together till the film is nearly ready to set. The following I have found to make fine slides in this way:—Cinnammic acid (gentle heat, *very* thin); succinic acid (rather strong heat, very thin); cinchonine (moderate heat and thickness); santonine (fair thickness). This last gives very various effects, some slides

FIG. 163.—Spring Wire Forceps.

showing a rough "ferny" pattern, while others are smooth in appearance. The most brilliant of all in colour, however, is benzoic acid, and it is also the easiest. It usually crystallises in long **straight** crystals (Frontispiece, B); but with a thinner film and very flat **glasses I** have obtained exquisite "**ferns.**" [1]

[1] Micro-crystallisations, as a **rule, do not** answer for the lantern. The microscopist desires perfect crystals **on a** small scale, whereas, on the screen, these would hardly show. For the screen, or the low power of a table polariscope, we want a much larger "**pattern,**" and brilliant **colour,** even at the expense of **what a** microscopist would call coarseness, but which on the screen **does** not appear **so.** Many beautiful crystals are available for the microscope which **are** useless for the table polariscope or the lantern, such for instance **as the** platino-cyanides. Microscopic crystallisations can be purchased in immense variety **for a shilling each.** Quinate of quinine and quinate of lime deserve special

Benzoic acid melted is especially convenient for a beautiful experiment. It melts at very moderate heat; and by preparing a wooden frame into which the double glass plate can be slipped and secured by a dovetailed slide while hot, crystallisation may be shown *proceeding upon the screen;* and the effect as the crystals shoot out in the most gorgeous colours is exceedingly beautiful. Cinnamic acid and the others crystallise with the same facility, but it is difficult to get the right thickness to show colour by this method, while benzoic acid never fails. Another attractive experiment of this sort is to place between two warm glasses a saturated *boiling* solution of silver nitrate. As this cools crystals form. The same slide when heated will re-dissolve the crystals, and is then ready to repeat the experiment. A correspondent informs me that a strong solution of urea also answers well for showing the process of crystallisation in colours upon the screen.

Many mineral sections show fine phenomena. Granite shows very well. Agate, if properly cut, polarises beautifully. Perthite is another good section. The prettiest I have seen, or possess, is a section of Zeolite, one of the hydrates of lime (Frontispiece, C).

140. **Organic Films.**—Most *organic* structures, having a decided "grain," are more elastic in one direction than in that at right angles to it. Hence, such as are transparent enough are doubly refracting, as wood would no doubt be if it were clear. As it is, some sections will polarise, but most of them are too opaque. The "grain" must, of course, for reasons we have seen, be placed at about 45° angle with the planes of polarisation to give the best effect. A quill pen thus placed in the stage shows beautiful fringes of colour; but it is better to cut off the barrels of two pens,

mention as exquisite micro-crystals. Many of these can be shown by the micro-attachment described on page 250.

slit them both up one side, and boil for an hour or two. They then become quite soft, and can be rolled out and dried flat between two of the glass discs, which they nearly cover, and make a fine slide. So does a sheet of horn, such as is used for stable lanterns, and which can be bought for a few pence. So will a piece of bladder, if of the right thickness to come within the colour-limits; or a few large fish-scales or gill-plates. Shrimp or prawn shells, soaked in turpentine and then mounted in balsam, polarise well. If the thickness is not quite right for colour, this can be brought out by a superposition film. When any one is known to care for such things, objects sometimes "turn up"; for instance, a friend once made me a present of a large human nail, placed at his disposal by an accident. This, boiled and flattened and scraped to the right thickness, made an interesting slide.

141. **Effects of Strain or Tension.**—If double refraction and its consequences be really due to unequal elasticities, we can readily demonstrate the fact by subjecting to unequal stress substances which in their natural state are homogeneous, and therefore show no double refraction. Thus, annealed glass, being of equal elasticity in all directions, does not polarise; but by making the elasticity unequal, it can easily be made to do so. Make a brass frame like Fig. 164, the size of the wooden slides, with a square-headed screw by which, with a T-handled key, strong pressure can be brought to bear through the centre in one direction,[1] the glass abutting against a convexity, A, opposite the screw, and

---

[1] Such pressure-frames are usually made by opticians with a thumbscrew, and flanges to keep the glass in place. But the chromatic effects heighten with the pressure; and as the glasses only cost about 2d. each, I prefer the T-key, and "put on" all I can, usually till I break the glass. The effects are far finer with this extra power, and best just before the smash comes.

being protected from the grinding of the screw by a convex padding, C, of brass or copper. At the least touch of the screw double refraction is shown by fringes of colour, and as the strain increases the effect is gorgeous beyond description (Plate IV. H). Then "unscrew" the objective a little in the collar, so that the glass and frame can be turned to 45° angle with the planes of polariser and analyser. The effect now is totally different, but equally beautiful. Next put in two bits

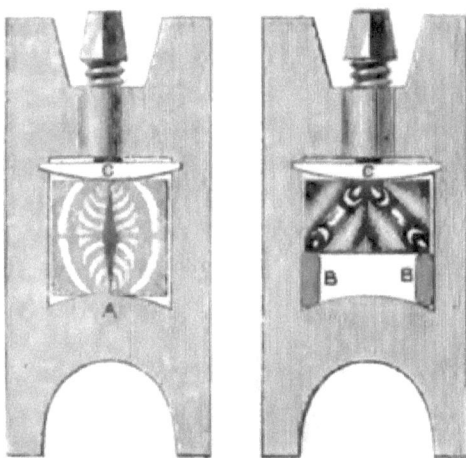

FIG. 164 —Screw Press for Glass.

of copper or brass, B B, in the *corners* of the frame opposite the screw, and abutting against them an oblong piece of glass. The strain now is different, resembling the breaking-load of a bridge. It will be seen how the coloured figures also differ, and how exactly the "lines of strain" are optically represented on the screen.

Other transparent substances will give the same effects. A glass trough, made the size of a slide, open at one *end* and filled with clear cold jelly, will show beautiful

T 2

phenomena if a rectangular piston is pushed in at the end so as to compress it. So will a "glycerine jujube," if compressed in any manner; but a still better plan, if a slab from which the lozenges are cut can be obtained, is to tie back the studs of the optical-stage, pass the slab of elastic matter through, and *extend* it with the two hands, of course at an angle of 45° with the polariser and analyser. A strip of thin transparent india-rubber will show similar phenomena when stretched. If neither is handy, soak some gelatine in cold water for a few hours, and then melt it with about two-thirds its weight of glycerine, and pour out upon a smooth stone or iron slab, greased, to cool. The composition will be something like that printers use for their rollers, but clearer; and an oblong slab passed through the slide-stage (kept clear by tying back the studs), and stretched, gives beautiful colour phenomena. Not much time must be wasted over such compositions, or jelly, or they will melt with the heat of the lantern, unless this is absorbed by an alum cell.

Heat applied to glass produces the same effects, owing to its expansive powers. Even one of the plain glass discs, fixed with a spring wire in one of the frames and held momentarily on alternate sides (so as not to crack it), with its centre over a small spirit-flame, will show a black cross, and transmit light through the rest when the analyser is crossed. But much better effects than this can be obtained. Make a "shell" of sheet-iron, like A B, Fig. 165, with a square hole in each side, $1\frac{1}{4}$ inches square, the parallelogram measuring 4 by $2\frac{1}{4}$ inches, so as to go in the slide-stage. A little bit turned over from top and bottom at A A, one end, makes a "stop" for adjustment. Cut a piece of wood, C C, such size and shape that the edge of one of the thick glasses made for the press just described "jams" in the shallow notch, and when wood and all are pushed in against the end stop, A, the glass stands central with the apertures. Fit a small bar of iron so as to slide in over the top edge of the glass. Having adjusted all except the bar of iron in the stage, and made the iron a dull red heat, slide it in over the

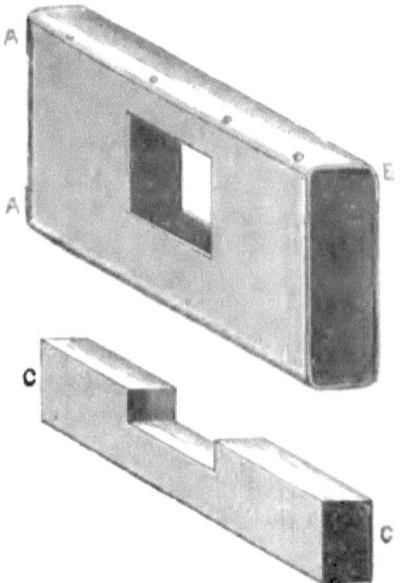

FIG. 165.—Apparatus for Heating Glass.

glass; at once fringes of light and darkness, and presently of colour, spread over the screen.[1]

---

[1] The private student needs no expensive apparatus for this class of experiments. As a boy, a quarter of a century ago, I first made the above experiment in the following manner. A dinner plate was inverted on a bare polished mahogany table, and on this was laid a rather massive square bar of heated iron. On this was "stood on edge" a 2-inch square of plate glass, its edges ground flat on a stone with sand and water. The table acted as polariser, and a few slips of glass placed in the card case of a medicine-bottle as analyser. I do not think I have ever experienced such pleasure from any experiment since.

In glass made red-hot and *suddenly* cooled, these beautiful effects are permanent. Such are called chilled or unannealed glasses, and cost from 4*s.* to 7*s.* 6*d.* each, of various sizes and patterns. In making them the great thing is to cool rapidly round the *edges*, and to start with a red heat. A square block made red-hot, and stood on one edge on an iron smooth surface, while any mass of smooth metal is balanced on the top edge, will show very good phenomena, especially if slid on to fresh cold surfaces till cold (Plate IV. G). A good chilled glass gives coloured figures particularly vivid, and the figures can always be foretold, a circular disc giving gorgeous coloured rings with a black cross. A good thickness ($\frac{1}{2}$ inch is not too much) does best for the lantern.

Owing to this property of glass in a state of tension, the polariscope is a most sensitive test of such a condition. One of the "Rupert's drops" or a Bologna flask will show conspicuous phenomena; and a correspondent once informed me that, having to renew the focusing lens of his polariscope, he found two in succession so powerfully doubly-refracting that it was impossible to get a dark field. It is in this way that optical glass is tested for valuable instruments, and it is by far the most sensitive test available.

142. **Effects of Sonorous Vibration.**—By this time we have a very vivid idea of the subtle power of polarised light as a revealer of the inner structure or molecular condition of bodies, provided they are transparent enough to apply it. The slightest difference in elasticity, or density, or, in short, from a homogeneous condition, at once stands revealed before this wonderful test. A singularly beautiful experiment, which Daguin describes as first made by Biot, though Dr. Tyndall first exhibited it by the lantern to a public audience, will show this power in a still more striking light. Get a strip of plate glass 5 feet to 6 feet long, 2 inches wide, and about $\frac{1}{4}$ inch

thick, A B, and smooth the sharp edges with a stone, or
with a file and turpentine. Prepare for it a wooden vice,
C, fitting into one of the wooden stands so often used, and
thus adjustable for height; and in this let the exact centre
of the strip be fixed[1] at an angle of 45° with the horizon,
or plane of polarisation.[2] Unscrew the slide-stage and
objective from the elbow of the polariscope, leaving only
the elbow, E, on the lantern, and support the "front" on
a wooden cradle of some sort (easily made by cutting semi-
circles out of the ends of a cigar-box) in its proper position
axially, but leaving a clear space of an inch or so between the
parts which ordinarily screw together. Through this interval
pass the slanting glass strip, A B, and adjust the height, &c.,
so that the strip may cross the field as near the place held
in the vice as possible. The whole arrangement (except
that the cradle for the "front" is omitted for the sake of
clearness) is shown in Fig. 166. Cross the analysing Nicol,
N, to give a dark field, and throw a loose cloth on the
"front" so as to stop all scattered light as much as possible.
(Half the battle in *all* lantern experiments, especially with
an inferior illuminating power, is to avoid such scattered
light, and many operators lose much effect by not attending
to it.) Now take a wet *flannel* cloth (other kinds "bite"
the glass too much and drag the vice about), and enveloping
the lower end of the strip in it, rub it smartly up or down
with a long smooth sweep. A shrill but wonderfully clear
musical note sounds out, from the longitudinal vibrations
into which the glass is thrown, and at each note the dark

---

[1] It is well to glue on the inner side of each jaw of the vice a circular thin slab of cork, so as to give a good pinch without breaking the glass.

[2] If the polariser is a Nicol, it is in some respects more convenient to set the polariser and analyser at 45° with the horizon, when the glass can be horizontal, and pinched at right angles.

screen is illuminated! If now a "chilled" glass be placed in front in the optical slide-stage, and focused as usual, and the experiment be repeated, at each note a quite different *colour* appears; or if a selenite butterfly be

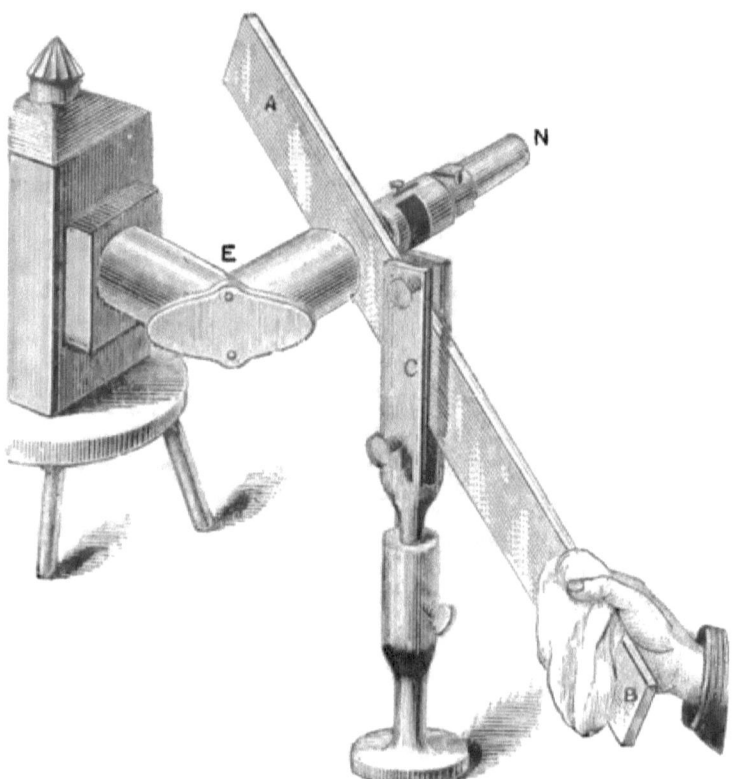

FIG. 166.- Effects of Sonorous Vibrations.

inserted, some other colour or that will appear. Or we may vary the experiment by putting in the arrangement for heating glass. Starting with a dark field, on inserting the hot iron bar we get the phenomena varied by the effects of *heat*-vibrations; and when we have got **good** colour, we vary these again by interposing *sound*-vibrations. This experiment can be very fairly performed with the Argand burner, and a screen distance of about 4½ feet, giving a disc of 15 inches.

This **beautiful** phenomenon is due to the stress caused in the **glass** near the nodal points, by the vibrations into which its molecules are thrown. Here, however, a difficulty may occur **to** some solitary student which actually did occur to myself, and which led me for **some** time to question this explanation. Dr. Tyndall himself states[1] that, upon sounding the glass, the screen effects are rendered "complementary"; and in my own experiments **I also** found this to be the case. The change from mere darkness to light only, is easily accounted for on the supposition that the thickness of glass, or the double refraction and consequent retardation, **are not** enough to produce colour; but when selenite designs give also "complementary" colours, the supposition naturally arises that the change of phenomena is of some *absolute* kind, and not one of *degree*, or comparison, as we should expect from a state of stress. Accordingly, I was for a considerable time inclined to attribute the phenomena to the half-wave retardation (§ 136) caused by the mere "resolution" of the plane-polarised ray, by the "absolute" motions of the glass molecules **at an** angle of 45°. But a valued correspondent[2] subsequently placed in my hands a translation of the researches of Kundt and Mach into this subject, which clear up the matter by showing that there *is* degree, or variable amount, in the effect produced; and that therefore the "complementary" results must depend upon the strip of glass giving an average retardation of about half a wave length.

[1] *Six Lectures on Light*, second edition, p. 139.
[2] The Rev. Philip R. Sleeman.

Kundt, having sent the light through the apparatus as described, analysed or spread into it a long band of light by a revolving mirror. This band was broken like a string of pearls, showing that the doubly-refracting effect was periodic, and coincident with the sonorous vibrations. Kundt then further interposed a selenite plate giving bright colour. The light being analysed as before by the revolving mirror, the band was found to *vary in colour*, the number of tints observable in the band increasing with the thickness of the glass or the intensity of the vibrations. Even thus, therefore, was established the degrees in doubly-refractive effect which the hypothesis of stress required.

But Mach carried the investigation still further by means of spectrum analysis. Selecting a selenite which gave at least two or three dark bands (§ 138), the light which passed through it and the glass bar was projected through a slit and prism as usual. When the bar was sounded, the dark bands became of course confused, and disappeared. But, assuming the slit and prism to be perpendicular, and the spectrum therefore horizontally dispersed, this spectrum was compressed into a narrower and more brilliant one by a cylindrical lens whose axis was horizontal, and then again dispersed vertically by a rotating mirror whose axis was also horizontal. Every colour and dark band was thus drawn out into a vertical string; and when all this was adjusted, the dark interference spots thus spectrally analysed separately at successive moments during the sounding of the glass were found drawn out into *zigzag curves*, whose amplitude represented the shifting of the bands, by the additional retarding effect of the temporary states of stress.

This beautiful experimental analysis places the true nature of the phenomena beyond any doubt.

CHAPTER XIII.

ROTATORY POLARISATION.

Phenomena of Quartz—Right and Left-handed Quartz—Circular Waves—Effect of Retardation in one such Wave—Rotation of the Plane of Polarisation—Use of a Bi-quartz—Rotation in Fluids—The Saccharometer—Chromatic Effects of Quartz—Other Rotatory Crystals—Electro-Magnetic Rotation—Important Difference between Magnetic and other Rotations—Connection between Rotation and Molecular Constitution.

143. **Phenomena of Quartz.**—We have seen that if we place in the slide-stage of the polariscope a plate of calcite cut perpendicularly to the optic axis, so that a beam of parallel plane-polarised light traverses it in the direction of that axis, there is no double refraction, but the plate behaves like a piece of annealed glass. But if we place in the same position a plate of quartz similarly cut, the effects are very different, though quartz also is a uni-axial crystal. The plate shows a beautiful *colour* all over; uniform if the light is nearly parallel and the plate truly cut; and as the analyser is rotated the colour passes in succession through more or less of the colours of the spectrum, never becoming white or black. That is, with white light. If we employ homogeneous light, as, for instance, a pure red, and adjust the analyser at $90°$ so as to give a perfectly dark field, on inserting the plate of quartz we find more or less of the

light is restored; but that on turning the analyser through a greater or less angle according to the thickness (for red light about 17° for every millimetre), the light is again cut off. It is as if the plane of polarisation from the polariser had been *rotated* in one direction or the other, the rotation being greater and greater as we proceed from red to the more refrangible rays.

144. Right and Left-handed Quartz.—It was soon found that some crystals required the analyser to be rotated to the *right*, and others to the *left*, to produce the same succession of colours; and Sir John Herschel found that this quality of right or left-handedness was connected with the "set" or direction of certain little facets often found on the shoulders of the crystals; thereby showing another connection between their optical properties and their molecular constitution.

145. Circular Waves.—Fresnel very soon came to the conclusion that the ray of plane-polarised light, on entering the quartz-plate parallel with its axis, was divided into two rays moving in *circular* orbits, of which one was more retarded than the other; and by marvellously ingenious reasoning, he further concluded that if this were so, he ought to obtain each ray separately by a compound prism, as in **Fig. 167**, where c is a prism of, say, right-hand quartz, and A and B half-prisms of the other, all cut so that the ray traverses them in the direction of the optic axis. He found that he did get two images, R^1 and R^2, in this way, which showed no variation in brightness as the analyser was rotated, and yet which differed from common light in giving colour when passed through a film of mica or selenite before reaching the analyser. Having verified this, the phenomena were easily explained. The original plane-polarised ray A B (Fig. 168), on entering the quartz, is divided into two rays moving in contrary *circular* orbits, whose directions are denoted by the double arrow-heads, R. If both moved with equal velocity, they could only meet together at the same point, A, of the circular orbit, or at the opposite end, B, of the diameter; but owing to one wave gaining on the other, the meeting will take place at some other

Fig. 167.—Fresnel's Quartz Prism.

point, R, depending on the thickness of quartz, and comparative velocities, and consequent amount of retardation. Now a circular orbit is compounded, as is well known, of a tangential force equal to a radial force at right angles to it. On emergence, therefore, it is plain the two tangential forces, S R, T R, will neutralise each other; thus leaving only the

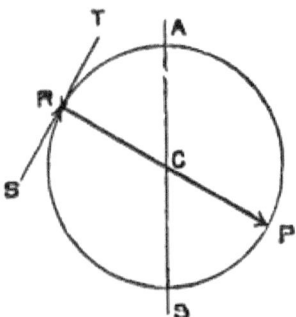

Fig. 168.—Rotation in Quartz.

two radial forces, R C, uniting in the plane wave, R P, which therefore represents the new plane of polarisation of the emergent ray. We know that the analyser must be turned at right angles to that to cut it off; and if *pure* red, or any other homogeneous light, be used, it does so; but since each wave-length is differently retarded, as usual, the analyser must be in a different position to extinguish each colour. We thus account for the successive colours, each of which is a compound tint composed of all those spectral colours *not extinguished* by the analyser in any given position.

146. **Proof by Spectrum Analysis.**—That the colours really are extinguished in succession, and the colours we see residual or compound colours, we easily prove by placing in the slide-stage a slit with our **quartz-plate**, and throwing its spectrum on the screen through the analyser; then we see the dark band or bands, which are extinguished in any given position, travel along the spectrum as the analyser is rotated. If we use a plate constructed half of left-handed quartz and half of right-handed, of the same thickness (called a bi-quartz), and place it so that the slit crosses both, these bands will travel in *opposite directions*, appearing in complementary positions in the spectrum; or yet again, placing the slit in the stage with a single quartz-plate, but with a double-image prism as **analyser, and** bringing the two images of the slit perpendicular, **on** rotating the Nicol polariser it will be seen that while the colours change in succession as before, the bands are always complementary in the two images. Hence, if we place in the stage with the quartz a circular aperture, and use the double-image prism as analyser (without the dispersion-prism) so as to produce two overlapping circular discs only, on rotating the analyser the colours change as before, but one disc is always complementary to the other, and the two make white where they overlap.

147. **Use of a Bi-quartz.**—If two quartzes of opposite rotations and of equal thickness be superposed in perfectly parallel plane-polarised rays,[1] they exactly counteract each

---

[1] The phenomena of convergent rays are described in a subsequent chapter.

other; and if unequal, the thinner one subtracts from the rotation of the other. As in addition to this it is obvious that the two halves of a "bi-quartz" must show the same colours in two positions of the analyser, while on the least rotation from these positions the colour of each half changes in *opposite directions*, it might be expected that a bi-quartz would be a most sensitive test of any *additional* rotary power added to either half. It is so: when both halves show the same colour, the addition of the merest film of either "right" or "left" makes **at once** a marked **difference**.

148. **Rotation in Fluids.—The Saccharometer.**—This would be of no particular utility were it not that many **fluids,** and amongst these *sugar syrup*, have the same rotary power very strongly; and as the degree of rotation is found to be strictly proportional to the amount of sugar, &c., in solution, as well as to the length of the column, we have here a most valuable means of ascertaining the strength of syrup.[1] Various arrangements are in use, some more sensitive than others; but nearly all depend on ascertaining either the degree of rotation, or the thickness of quartz, necessary to balance the difference of colour or other phenomena caused by a column of solution **or** fluid of **a** standard length. This length is very much greater than the thickness of quartz

---

[1] Till very lately there has been no really scientific work accessible in English upon the use of the polariscope in practical chemistry. As a consequence, the methods employed have been often somewhat rough, **and** necessary corrections and precautions were neglected. The recent publication of a translation from Dr. Landolt's classical treatise (*Handbook of the Polariscope, and its Practical Applications*: Macmillan and Co.) has supplied this want, and placed the analytical chemist in possession of all he needs to know. The principal forms of chemical polariscopes are described and figured in this work; but I do not think Jellett's form of double-prism (for description of which see Reports of the British Association, 1860, ii. 13) has nearly had justice done to it in comparison with others; in fact it has often struck me that Germans rarely do justice to English physicists or inventors.

necessary to produce a given rotation; hence it is usual to employ tubes with plane glass ends, from 6 to 24 inches in length. For mere experimental demonstration, a brass or glass tube 2 inches diameter and 3 to 6 inches long, may be capped with glass at each end, and filled with saturated sugar syrup. The front and objective being unscrewed from the reflecting elbow of the polariscope, and supported a little in front as for the experiment with the strip of glass, the bi-quartz may be placed in the stage, focused, and its two colours equalised. If the tube with the sugar be then supported or held by hand between the polariser and the quartz, so that the polarised beam has to traverse it, it will be seen that the colour of the two halves of the quartz plate is at once differentiated, and that the analyser has to be perceptibly rotated to bring them both alike again. Spirits of turpentine give the same phenomena, and oil of lemons in a still more marked degree. By using more sensitive bi-quartzes, in which each half is composed of superposed *wedges* of right and left-handed, or of slices cut in other directions and put together in particular ways, displacement of the equal colours may be shown on the screen by a cell containing only 1 inch of fluid, which will go with the quartz into the optical stage. One of the best forms is that shown in Fig. 169, where A is a wedge, say, of right-handed, and B of left-handed quartz in the half A B, whereas in the other half C is left-handed and D right-handed. Across the middle, where all four wedges are of equal thickness, there is, of course, a black line when the analyser is crossed, and white when it is parallel with the polariser, the right and left quartzes there neutralising each other. On each side of this there are necessarily coloured straight bands, as shown, exactly similar in two positions of the analyser; but as in one half the right-hand quartz is thickest in all below the middle line, and in the other half the left-hand quartz, any rotation of the analyser, or introduction of fluid or substance with rotary power, displaces all the bands in opposite directions, and a very slight addition of rotary power is thus perceptible.[1]

The variety of effect possible from various wedges or sections of quartz in combination, or rotated, is in fact almost endless; but it may be mentioned that a large concave plate cut across the axis gives, of course, Newton's rings, whose colours change, the rings moving inwards or

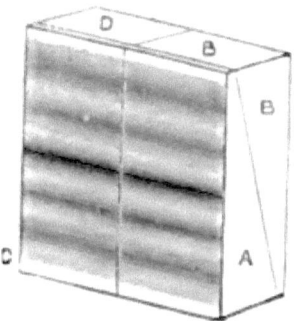

FIG. 169.—Compound Wedge Bi-quartz.

outwards as the analyser is rotated. The effect of this is very beautiful; but it will be shown later on (§ 156) how the same appearances may be produced with a concave selenite.

[1] In Mitscherlich's instrument, there is no test-quartz or other apparatus, but the amount of rotation needed to restore a simple dark field after the column of fluid is introduced, is observed by the light of a sodium flame. In Soleil's a bi-quartz is used. In Wild's a double quartz-plate showing "Savart's bands" (see § 179) is employed, the bands being first extinguished, and the angular rotation necessary to restore extinction being observed. In Jellett's instrument a peculiar calc-spar prism is the analyser, which gives two images polarised in two planes very slightly inclined, and which are brought to equal brightness: then the rotary substance differentiates them. Cornu employs in the same way, but as polariser, a Nicol cut longitudinally in two, which halves are very slightly inclined to each other, and so also give two inclined planes of polarisation.

Lastly, a large quartz plate is exceedingly good as a "superposition" plate, as the colour may be changed by rotation of the analyser till the most pleasing effect is produced. It is somewhat strange that either quartz plates, or substitutes for them constructed of two mica films as hereafter described, are not more used for microscopic superposition work, instead of a complicated set of films in expensive mountings.

149. Other Rotatory Crystals.—Some other crystals besides quartz have the power of rotating a polarised beam. Amongst these may be mentioned cinnabar, periodate of soda, sulphate of strychnia; and there are a few others. Chloride of sodium (common salt) has sometimes the property *in all directions*, therein resembling fluids.

150. Electro-magnetic Rotation.—Faraday discovered that a cylinder of glass placed with its ends axially between two poles of a powerful magnet, or surrounded by a helix traversed by an electric current, similarly rotated the plane of polarisation, differentiating the two colours of a bi-quartz. His "heavy" glass (borate of lead), and some peculiar forms of flint-glass, give the most effect; good common flint having a rotatory power about half that of the "heavy" glass. Further experiments have shown that the electro-magnetic force rotates the beam in some degree through nearly all, if not all, liquids, and even gases, thus opening up many most interesting questions of molecular physics. Dr. Kerr further discovered that a beam of plane-polarised light reflected from the polished pole of an electro-magnet, also showed effects of rotation whenever the iron was magnetised by the passage of a current.

But there is this remarkable difference between electro-magnetic rotation and other forms of the same phenomenon. If we send the ray through a quartz which rotates it to the left of the observer, and then reverse the quartz, the ray is still rotated in the same direction. Consequently, if we reflect the ray back again through the same quartz, the rotation conferred by the first transmission is exactly reversed by the second: the circular waves *go back* upon their former paths. In glass or fluid whose rotary property depends upon electro-magnetic action, on the contrary, the rotation is repeated in a reflected ray; and whichever way the ray is transmitted, and as often as it is transmitted, follows the direction of the current.

151. **Rotation and Molecular Constitution.**—We have seen that a limited number of crystals possess rotatory power. Nearly all of them are found both right-handed and left-handed. They are all either uni-axial or have no double-refraction at all. And, as a rule, if not universally,[1] those of them which are soluble in any fluid, lose the property of rotation when so dissolved. We are driven to the conclusion, that in the case of crystals the property of rotation is dependent upon crystalline structure; or, in other words, upon the arrangement of *groups* of molecules. This supposition is confirmed by the fact that in most of such substances the ray of light must pass through them in one particular direction—that of the single optic axis—to **be** rotated. And it is still more strongly confirmed by the successful production of rotatory mica-combinations as hereafter described (§ 157).

We might suppose the same of fluids—that is, that the property depended upon the arrangement of molecular *groups*—were rotation confined to the fluid state. But experiment proves that when rotary fluids are volatilised, the *vapour* exerts sensibly the same rotatory power, for equal weights, as the fluid. We are, therefore, forced to the conclusion that in this case the property depends upon the *single* molecules of which the gaseous matter is composed. As analogy compels us still to regard it as dependent upon

---

[1] I am not sure that the rule is quite universal.

some "structure," we are **driven to the** further deduction that here rotation depends upon the grouping **of** *atoms* into the molecule. It has, indeed, been disputed whether in fluids there really is any real *circular double-refraction*, **as** in quartz, whose two circular waves were so triumphantly separated **by** Fresnel; and Dove's experiments failed **to** decide the matter.[1] But the primary phenomena which led Fresnel to *suppose* it in quartz exist in the case of fluids; he himself, in a paper presented to the French Academy on March 30, 1818, reports experiments which he held to bear out this view; and my own experiments in the production **of** spiral figures (see Chapter XVI.), in which quartz is successfully replaced by either rotatory fluids or mica combinations, are very strong evidence in favour of the existence of similar circular waves.

Regarding the molecule, then, as the source of **the** property, it is surely strangely significant to find that all, **or** nearly all, the substances possessing rotatory power in solution or vapour are ***complex carbon*** *compounds*. Many beautiful and refined researches by Pasteur, Van't Hoff, and others, have moreover shown that in all such rotatory substances **there** are either one or more *asymmetrical* carbon-atoms;[2] **but** on the other hand, there are many substances which **contain** such asymmetrical atoms **in** which no rotation has **yet** been observed. Against this, however, is to be set the remarkable fact that a tartaric acid with no rotation has been separated into two **varieties** of *opposite rotations*, whose combination makes **the** neutral form. These two forms of rotatory tartaric **acid, though** in most reactions and in chemical constitution **identical**, differ in some reactions,

[1] Poggendorf's *Annalen*, **cx. 290**.

[2] I believe this statement correctly represents our knowledge of the matter at the time it is written. Should any exceptions have been discovered, **or** in future be so, these will not much affect the physical **aspect** of the fact as a general rule.

and crystallise in right-handed and left-handed forms (§ 144). There are a few other substances in which **the inactive** and one rotational form, but not the other, are known; and some —especially several ethereal oils—in which opposite rotations, but not the neutral form, are known. Some of these forms have been obtained with great difficulty; and hence the theory has been advanced, and **is** highly probable, that all carbon compounds containing asymmetrical atoms really are composed of molecules whose atoms **are** arranged helically, but that in the inactive ones two such molecules of opposite rotations are combined in a **kind of sexual** combination, and have not yet been separated.

It is further to be observed, that substances which possess rotatory power in fluid or solution, may often be varied or reversed in power by the addition of other fluids or substances. And it is still more remarkable, that when capable of crystallisation, they crystallise almost, if not quite, invariably as *bi-axials* (see **Chapter** XV.), which itself is a strong evidence (see § 170) of unsymmetrical atomic structure. **In** crystallising **they** appear to lose their rotatory power; but as this would be, with our present means of observation, overpowered in any direction of the ray by the far stronger ordinary double refraction, this can hardly be held to be proved. If the crystals are melted, however, so as to destroy the crystalline structure and restore the amorphous condition of a fluid—as, if a crystal of sugar be fused—the rotatory property remains **or** is restored. Heat will also convert some rotatory substances into inactive forms; as if there were a tendency under the influence of heat vibrations for a number of molecules to pass into the form of the opposite *sex*, as it were; the two kinds afterwards uniting in pairs. But enough has been said to show the exceeding interest of the subject, and the light it is likely to throw upon the nature of atomic grouping and the structure of molecules.

CHAPTER XIV.

CIRCULAR AND ELLIPTIC POLARISATION.

Fresnel's Rhomb—Composition of **two** Rectilinear Vibrations into a Circular one—Quarter-Wave Plates—Other Methods of Producing Circular Polarisation—Rotational Colours of Circularly-Polarised Light—Reusch's Artificial Quartzes—Behaviour of Quartz in Circular **Light—Phenomena of** Thin Films when Analysed **as well as Polarised Circularly.**

152. **Fresnel's Rhomb.**—We have thus examined the effects of contrary circular orbits as exhibited in quartz; but circular orbits, or circular polarisation, may be produced in other ways, which furnish **a** most conclusive and elegant **proof** of the truth of **the** Undulatory Theory, showing as **they do how** the motions we **are** dealing with **answer to every process** to which they can be subjected. Fresnel not **only tested his theory of the phenomena by** separating the **two actual rays in quartz; but** ·he further calculated from **his mathematical conceptions, that if he** constructed a **rhomb of glass with parallel faces so** disposed that a **ray (A B, Fig. 170), was** "totally reflected" twice within it **at an angle of 54° 37′ as** shown; **if that ray was** plane-polarised in a plane *inclined at 45° to the reflecting surfaces*, **it would be as it were so** divided and spun round by the relation **of the reflecting** surfaces to its vibrations, **as to**

emerge circularly polarised; as at C.[1] This was entirely worked out first in theory; but experiment verified it. The ray *did* emerge circularly polarised; for rotation of the analyser showed no difference in brilliancy, and yet it differed from common light in causing colour, when passed before analysation through doubly-refracting films. That was a great triumph for the theory; but there was soon another, and one more easily intelligible to those acquainted with only elementary mechanics.

153. **Composition of Vibrations.**—A moment's reflection showed us that two plane-polarised rays vibrating in the planes — and | , even though originally from the same

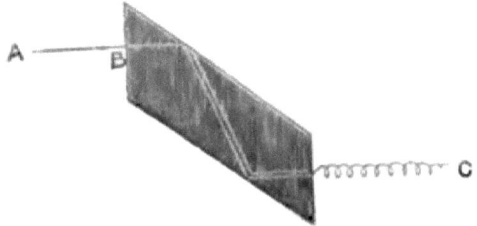

FIG. 170.—Fresnel's Rhomb.

polarised ray, cannot possibly interfere with or quench each other. In *that* way, they can **have no** relations. But yet they may act on each other in another way. Let a pendulum be mounted as in Fig. **171, so** that it swings on gymbals, G, from two **axes at** right angles to each other;[2] if swung on one axis the bob will vibrate **in** the path A B; if on the other, in the path C D at right angles with it. Let these

[1] I give this **as** conveying a rough, popular idea; in reality, the original ray is divided into two at the first reflection, of which one is retarded ⅛ of a wave-length, and still more differentiated to a quarter-vibration by the second reflection. The rest follows as in the next method. See for details Lloyd's Lectures on the Wave Theory.

[2] The illustration in this form is, I believe, first due to Professor Baden-Powell.

represent the two planes of plane polarisation, and the bob a molecule of ether, and let it have arrived at B in the plane orbit A B. It has therefore reached the limit of its swing, and the next moment will begin to swing back, but at this moment has no motion. Just at that moment, then,

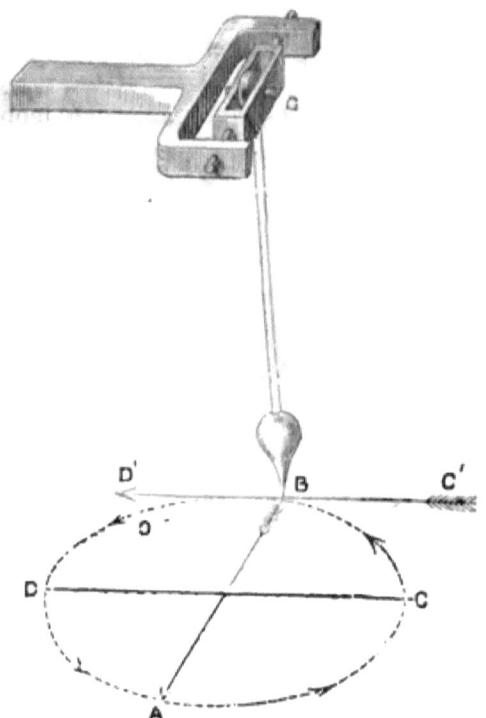

FIG. 171.—Composition of a Circular Vibration.

imagine the bob of a duplicate pendulum, moving in an orbit at right angles to A B, and in the exact middle, or full power of its swing, to strike against it, as represented by the arrow C' D'. This second bob will yield up its motion and come to rest, and may be withdrawn; but its transferred force, thus applied tangentially, will be compounded with the other, and drive the bob, B, into a new *circular orbit*, B O D, in the direction of the arrow.

Now we **see** at a glance that the second vibration, to do this, must be exactly *a quarter of a whole or double vibration* before or behind the other, or must be at its fullest power when the other is at the moment of rest; and it follows, therefore, that supposing us to be dealing with actual physical realities—real atoms of ether vibrating in real paths—a circularly-polarised ray *ought* to result, if we caused one plane-polarised beam to be retarded exactly a quarter of a wave, **or** any odd number of quarter-waves, behind the other. If, again, retarded somewhat differently from a quarter-wave interval, experiment with the pendulums gives an ellipse, and this also ought to result. And, lastly, it is easy to see that if the **two** plane rectangular vibrations differ *half* a vibration in phase, they must compound together one *plane* vibration at an angle of 45° with each of them, but in which of two rectangular directions depends upon which is most retarded.[1]

154. **Quarter-Wave Plates.**—Such being the theoretical view of the matter, we have already learnt that in several ways we can accomplish such a result; but the simplest are the use of compressed glass (due to Dove), or of a thin film of selenite or mica, called a quarter-wave or quarter-undulation plate (due to Airy). It is exceedingly difficult to get a large even film sufficiently thin in selenite; and therefore mica is usually employed, with which it is very easy to procure one, though the thickness is very small. First cut a rather thin film of good *even* mica, exactly the size of the glass discs between which it will be mounted. Then stick a very sharp but not too flimsy needle by the eye in a

[1] The student is strongly recommended to work all these cases out by diagram.

handle, and with the point split the mica as thin as possible. After that, by rubbing the smooth *side* of the needle gently along the edge of the mica, this will be bruised rather thicker, so that it can be split thinner still, taking care to split gently and evenly, and rather *wedging* the films gently apart by coaxing the needle carefully inwards. It is well to split a dozen or so in this way, as thin as possible, and taking care not to soil them. Then have a few perfectly clean glass discs, place each film between a pair of the glass discs for handling, and test them in the polariscope used as a table instrument. We shall find one or two films, unless we are very unlucky, which when placed in the stage in the position that would give colour were it thicker (they are always tested, and used, in the rotating frame), give always the same illumination as the analyser is rotated, the only variation in colour being from a rather *bluish* grey to a rather yellow or *fawn-coloured* grey. The reason of this is, that *all* the colours cannot be exactly a quarter-wave different, simply because the wave-lengths vary; it is best, therefore, to work to the yellow as both the brightest colour and near the middle of the spectrum. The test of an equal light will be sufficient for most experiments; but a more sensitive and exact test is with the Norremberg "doubler." If the film is laid on the bottom mirror in the right position, the quarter-wave retardation being "doubled" into a half-wave, when the analyser is crossed and adds another half-wave (§ 136) the light ought to be almost fully transmitted; whilst with the analyser parallel the tint ought to be the first plum-colour or "transition-tint" between Newton's red of the first and blue of the second orders, showing extinction as nearly as the various wave-lengths of the different colours allow.

This then is our "quarter-wave" plate, which should be at once mounted between two glasses in balsam, and its working planes marked on the edges by scratching with a diamond, or quartz crystal. Meantime, however, observe how beautifully the phenomena correspond with theory; for, placing the plate in the optical stage in the rotating frame in the proper position, it will be seen that the double-image prism gives equal images in all positions; while yet we shall find, by beautiful phenomena, that the light is still "polarised," though in a different way.

155. **Plane and Elliptical Composition of Vibrations.**—The student will readily perceive it must follow from the foregoing, that with any uneven film much of the light transmitted must be either circularly or elliptically polarised; and that with any film, such as a Newton's ring slide, ground with regularly graded thicknesses, regularly recurring bands must be circularly polarised, **and** intermediate bands elliptically polarised and plane polarised. Experiment, by any of the methods hereafter to be described which distinguish circularly-polarised light, readily proves that such is really the case.

Light is also circularly or elliptically polarised by reflection from metals—almost circularly by silver.[1] **And it can also** be circularly polarised by placing our glass press in the stage, with the screw-pressure directed at 45° to the polarising planes. Then it is obvious that just enough stress can easily be given to circularly polarise the light, as this effect solely depends on a *given amount* of retardation. By this method, however, only a space in the centre of the glass— about one-fourth of the whole surface—is at all uniform, the outer edges being too strongly doubly-refracting, owing to the greater stress in those regions. Mica films are, however, best, being so easily prepared, and giving the entire field. A second smaller quarter-wave film should also be mounted

[1] Light is not plane-polarised by silver at any angle, and circularly only at a certain angle. Hence we see why the silvered surface of the "thin film," in § 100 did not quench the light.

between glass discs, **and set in** a short bit of brass tube, or a narrow edging of **cork**, by either of which it can be inserted at pleasure into the end, B, of the **crystal** stage shown in Fig. 181. Thus equipped we are ready for a further most interesting set of experiments. Before commencing them, however, we must clearly understand the positions of the apparatus. Representing a quarter-wave plate by Fig. 172, A B and C D are its planes of vibration or polarisation, at angles of 45° with the supposed planes of polariser and analyser. This is the *normal* position of the plate, in which it should be marked on the edges at E F, G H; it is also the usual position of a coloured film. But whenever we

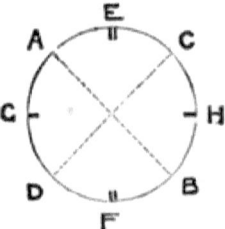

FIG. 172.—Quarter-Wave Plate.

describe the quarter-wave plate as "at an angle of 45°," it **is** meant that the *marks* stand diagonally, the planes of vibration being then vertical and horizontal, or corresponding with those of polariser and analyser.

156. **Rotational Colours of Circularly-polarised Light.**—Take now any film of mica or selenite, which gives good colours, and place it in the stage in the usual position. We have already found that it gives complementary colours with analyser crossed and parallel, and no colour at all with **analyser** at 45°. But now *after* the film, or between it and **the** analyser, introduce the quarter-wave plate at an angle **of 45°**. (This is not its usual position, as just explained,

and is rendered necessary by the fact that the colour-film has *already* changed the planes of vibration by that angle.) Observe the difference. As the analyser is turned the colours successively change, much like those of a plate of quartz cut across the axis, the order of the tints depending on the direction in which the analyser is rotated.

But it also depends on the relation of those planes **in** film and plate which transmit the ray of greatest retardation; **for,** having observed the order of succession in whichever position has happened to come first, rotate the quarter-wave plate in the frame (the plate should always be mounted between bare glass discs, and used in the rotator) till it stands **at** right angles with its first position. **The order** of colours for the same rotation of the analyser is **now reversed.**

But thirdly, if the film instead **be** rotated 90° we have the same effect; and this gives us a very beautiful modification of the experiment, which places the matter in a more striking light. From a plate **of** mica or **selenite,** which gives good colours, cut **a** square about one inch in each side, whose sides shall be perpendicular **(as in** Fig. 173) when the planes of polarisation are at 45° (the position **of** most brilliant colour). Cut it in two vertically by the line A B. A moment's inspection of the position of the polarising planes, as denoted by the dotted diagonals, shows that the *inversion*, either laterally or vertically, of one of the halves, is equivalent to rotating these planes 90°. Invert one-half accordingly,[1] and having put the two halves together again in the new position, mount in balsam between glasses. By plane-polarised light, these changes of

---

[1] A still more striking slide can be made by inscribing the square in the full circular field, and cutting through so as to divide the containing segments of the circle, and the square into four triangles by the diagonals. Then inverting the top and bottom quarters of the square, and the side segments of the circle, we get a geometrical figure in eight portions, of which every adjoining pair change colour in reverse order.

position make no difference whatever; the whole plate still gives the same colours, complementary in opposite positions of the analyser. Introduce the quarter-wave plate as before. When the analyser is either crossed or parallel the colour is still uniform, and complementary in the two positions; but between, as it is rotated, the colours change as before, *but in reverse order*, giving different colours like A B, Fig. 174. In fact, the slide gives us the same phenomena, nearly, as the bi-quartz in §§ 146, 147.

A selenite Newton's ring slide, placed in the same circumstances, gives the same successive colours; and hence exhibits the beautiful phenomenon of all the rings moving

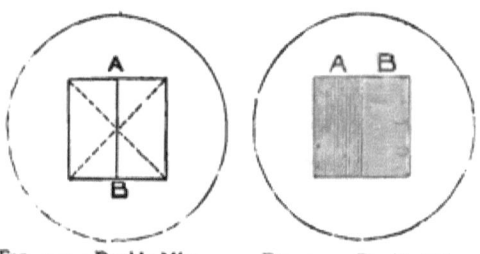

FIG. 173.—Double Mica.   FIG. 174.—Double Mica.

from or to the centre, somewhat like a chromatrope, according to the rotation of the analyser, and the respective positions of the principal plane in film and plate. Or if the quarter-wave plate be steadily rotated, for reasons already seen, the rings *alternately* expand and contract. All other selenite slides show the same effects; so that a chameleon, for instance, instead of giving only two complementary colours, passes in succession through nearly all the colours of the spectrum. Or if a quarter-wave plate made in two halves divided horizontally, and so cut that the principal planes are at right angles, be superposed on such a wedge as is shown in Plate IV. Fig. A, the colours will appear to pass along the wedge in opposite directions in the top and bottom bands as the analyser is rotated, the stripes only appearing single in two complementary positions.

These facts are capable of useful application in super-position films for the microscope. Instead of the complicated "selenite stages" mounted in brass, a single film judiciously chosen, mounted with a quarter-wave plate on the top, in proper position, between two protecting glasses, will give any colour the microscopist can desire, in one plain glass slide to be simply laid upon the stage of the instrument.

157. **Reusch's Artificial Quartzes.**—The resemblance of these rotational colours to the phenomena presented by quartz, has already been alluded to; but the resemblance so far is incomplete, the succession of colours with two films being neither so regular nor so perfect as with a quartz of sufficient thickness. But it occurred to Professor Reusch, who was acquainted with the artificial uni-axial crystals composed of crossed mica-films described in the next chapter, and discovered by Norremberg, his predecessor at the University of Tübingen, that by superposing thin films of mica with their similar polarising planes successively adjusted at regular angles round an axis, the phenomena of quartz should be *perfectly* reproduced. His method is shown in Figs. 175 and 176, which represent arrangements for a right- and left-handed combination. The three rectangular slips numbered 1, 2, 3 in each figure, are so cut that the same "principal" polarising plane (that which contains the two "optic axes" as explained in the next chapter) is parallel with their longest sides. They are then superposed in the order of the numerals, each slip at an angle of 60° with the one underneath; and in this way a tolerably large number of films (which must be a multiple

of three—say fifteen to thirty and very thin), are built up and cemented together with balsam and benzol. Such combinations will give a perfect rotation of colours in the direction of the numbers, right or left. Instead of an angle of 60°, we may adopt that of 45°, or half a right angle; in which case it will require four films to complete each circle, and the total number superposed must be a multiple of four.

These artificial quartzes offer a beautiful proof of the theory of Fresnel respecting the two contrary circular waves. Mere inspection of Fig. 175 will show how one of the

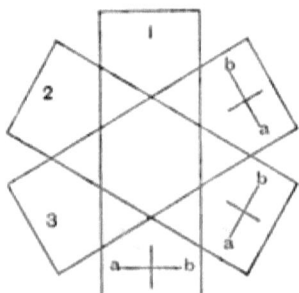

FIG. 175.—Mica Quartz.

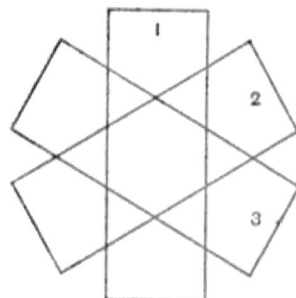

FIG. 176.—Mica Quartz. Reversed Rotation.

polarising planes, ab, is successively changed in direction; so that it is practically converted into a circular orbit, making several revolutions while the ray gets through the total combination. And detailed analysis of the "resolution" of the rectangular vibrations by each successive film, and finally by the analyser, will show that the total result is approximately equal to that of two *contrary* circular waves, of which one is more retarded than the other. Fresnel's theory was thus verified in detail, years after the author had been removed from the scene of his brilliant discoveries. These combinations moreover give, perfectly, *all* the phenomena of quartz as further described in the next chapter (§ 178).

Professor Reusch's films were actually built up as described, of oblong slips; but such is not the best way of making them in practice. The long slips consume so much mica, that it is difficult to get all the films exactly equal, which is essential. But this object can be easily obtained by another method of construction, which is that adopted by Mr. C. J. Fox, in making the finest preparations of the kind which have ever come under my notice. A largish slab of mica must be procured, from which can be split even films as thin and large as possible—say nine or ten inches long by five or six inches wide. On this must be carefully found by experiment, and scratched, a line to show the principal polarising planes. Then all the films for the same preparation are to be cut *from one and the same sheet*. As good a plan as any is to draw on a sheet of paper squares at the proper angles, and laying the mica upon it, scratch over the lines. For instance, supposing the preparation is to be built up of films at 45°, half the sheet will be covered by squares like Fig. 177, with sides parallel to the polarising planes, and half by squares at an angle of 45° with them, as Fig. 178. These may be cut with scissors; but special care must be taken that none are inverted or turned round, else all will be disarranged and all labour lost; hence it is best to scratch some mark in one given corner of every square, by which the position of that corner can be distinguished. Then if we take first a square from Fig. 177, and superpose on it a square from Fig. 178, *turned round* 45°, *say to the left, so as to coincide;* next another from Fig. 177, *turned round* 90° *to the left;* and next a second from Fig. 178, turned round 90° to the left; we shall have gone accurately round the circle, and the whole preparation can be built up in this manner. If 60° be the angle adopted, equilateral

x

triangles may be similarly treated, and the alternate triangles with apex inverted may be **used** for a preparation of contrary rotation. The best effects are obtained with not less than about 24 films, as near $\frac{1}{8}$ wave thick as possible ; but Mr. Fox has made **fine** preparations consisting of as many as 42 films ; **and on the** other hand, moderately good results may be got **from as** few as a dozen films **a** quarter-wave thick, arranged in four ternary **sets.** The one main thing is absolute uniformity in thickness, which can only be obtained by using the same sheet for all the successive films.

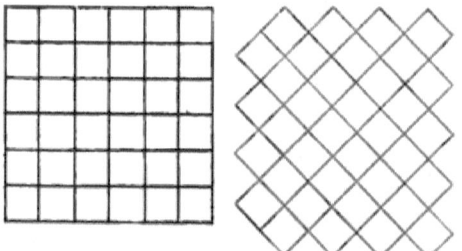

Fig. 177.—Mica Squares.   Fig. 178.—Mica Squares.

158. Quartz in Circularly-Polarised Light.—Con versely to these phenomena, if the quartz or bi-quartz **be placed** in the optical stage, **and the** quarter-wave plate introduced between **it and the** analyser (in this case in its **normal** position, **as the plane of** polarisation to which it is related is that **of** the polariser), **this** will be found to behave nearly like **a** film **of** mica, giving in the main two complementary **colours and** scarcely any intermediate colours. This slight qualification is necessary, because of the fact that **all** the **colours, as we have before seen,** cannot be *quite* circularly polarised by a plate of mica.

159. Further Chromatic Phenomena of Thin Films.—Still further, we know that if we rotate a film between **the** polariser and analyser, whenever its planes are

FURTHER PHENOMENA OF FILMS.

at 45°, it exhibits its colours ; but when the planes correspond with those of polariser and analyser, there is no colour—in fact no effect at all is produced, for obvious reasons already explained. But now place **first in** the stage the large quarter-wave plate in its normal position ; insert in the stage also a *second* quarter-wave plate with its principal plane at right angles to that of the first ;[1] and between them, or after the first plate in the stage, the double mica-film (Fig. 173), or the geometrical figure described in **the foot-note.** The circularly-polarised ray from **the first** plate, after being doubly-refracted by **the** film, is now again converted into a circularly-polarised ray. It has, therefore, **lost all** trace **of** sides or plane polarity ; **and when the** analyser **is crossed,** if the double film **is** also **in a rotating** frame, this **can** now be rotated *without the colours changing in the least.* **The** best way is to adjust it first at an angle of 45° **with the** usual position for a coloured film (a position which would show no colour at all but for the first quarter-wave plate). Then the crossed position of analyser may be found by the two tints being equalised ; and if then the bi-mica or compound selenite be rotated, the uniform colour will be preserved in all positions. **If then** again brought to the unusual position just indicated, and the *analyser* is rotated, we get the same successive, or bi-quartz phenomena, already described.

[1] The most convenient plan **is to** have a second smaller quarter-wave plate fitted in the crystal stage described further on (Fig. 181).

CHAPTER XV.

OPTICAL PHENOMENA OF CRYSTALS IN PLANE-POLARISED LIGHT.

Rings in **Uni-axial** Crystals—Cause of the Black Cross—**Apparatus for** Projection or Observation— Preparation of Crystals— Artificial Crystals—Anomalous Dispersion in Apophyllite Rings—Quartz—Bi-axial Crystals—Apparatus for Wide-angled Bi-axials—Anomalous Dispersion in Bi-axials—Fresnel's Theory of Bi-axials—Deductions from it—Mitscherlich's Experiment—Conical and Cylindrical Refraction—Relations of the Axes in Uni-axials and Bi-axials—Composite, Irregular, and **Hemitrope** Crystals — Mica **and Selenite** Combinations—Crossed Crystals—Norremberg's **Uni-axial Mica** Combinations—Airy's Spirals—Savart's Bands.

160. **Rings in Uni-axial Crystals.**—To some extent we have found proofs already of the close connection between the form and other physical characteristics of crystals, and the optical phenomena they present. But seeing we have found polarised light such a delicate analyser of the inner constitution of bodies, acting as a sure revealer of any state of unequal tension, however caused; of invisible sonorous vibrations; and even of electro-magnetic stress; it is natural to inquire if it will not yield us further evidence of the molecular constitution of the crystals themselves, or the plan on which they are, as it were, built up. This line of inquiry takes us into a new and magnificent range of

optical phenomena, to enter which we have simply to abandon the nearly *parallel* beam of light we have hitherto employed, bringing to bear upon our crystals pencils of rays distinctly convergent. Provide a plate of Iceland spar, cut across the axis, and about ⅛ inch thick. It need not be large; and for the lantern polariscope it may be conveniently mounted between two thin glass circles, like Fig. 179, in the centre of a wooden slide 4 inches long by 1⅛ inches wide. Placed in the centre of the optical stage, so as to get the *parallel* beam of polarised light, we have already discovered (§ 121) that it acts as a plate of glass would do, producing no effect; its image is dark or light, according as the analyser is crossed or parallel. But now imagine *converging* or

Fig. 179.—Calcite Plate in Slider.

diverging rays, or a conical beam of plane-polarised rays, such as are given by a lens, passing through the slice; it is evident that only the central rays can pass exactly along the optic axis; and hence inclined rays must be more or less doubly refracted. At equal distances all round the centre, therefore, the slice must act as a thin film, and give colour arranged in symmetrical circles. But this is not all. We have already learnt the two directions into which the original polarised plane of vibration must be resolved in the crystal, and that one of the new planes must *pass through the axis*, while the other of course is at right angles to it (§ 122). Taking therefore Fig. 180 as a diagram of our slice, and supposing A B to be the original plane of vibration from the polariser, the plate of calcite resolves that, everywhere, into vibrations passing through the axis, represented by the radii; and others at right angles to them, represented by the circles. Further still, it is evident that all along the two radii, A B and C D, there will be no double refraction at all, since the planes of vibration in the crystal there coincide with that of the polariser, and that perpendicular to it. Along those lines, therefore, the slice can have no influence, and must appear black when the analyser is crossed, and white when it is parallel with the polariser (§ 136).

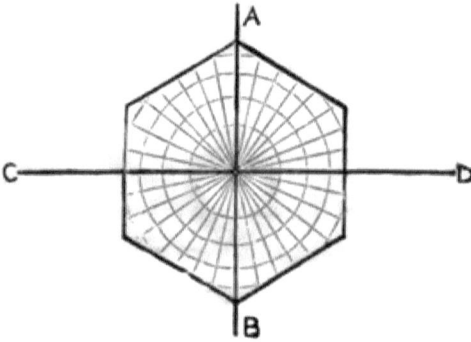

FIG. 180.—Vibration Planes in the Calcite.

All this is justified to the letter; but to exhibit it on the screen, we must add to the polariscope what is called a "crystal stage," which will hold the plate in the converging cone of rays from the objective. Fig. 181 shows the construction of it. A tube, A B, fits on the nozzle of the objective, and has transversely cut through it a slit, S, through which the crystal sliders are passed, kept in place by a spiral spring as usual. The end, B, of the tube is of exactly the same size as the nozzle, so that the Nicol or other analyser fits and rotates on it, as on the nozzle. We place this addition on the nozzle, add the analyser, and insert our plate of calcite in the slit, S. We then get on the screen the beautiful figures represented at A and B, Plate **V.**, according as the analyser is crossed, or parallel with the polariser. In the former position the beautiful coloured **rings are** traversed by the black cross we **were** led to expect; **in** the other position, we get the complementary rings traversed by a white cross. The centre, of course, shows no phenomena at all beyond white or black, as the rays there pass along the optic axis.

161. **Apparatus for Observing the Rings.**—The objective described in Fig. 1 gives, in practical work, about the best average effects with these "crystal rings," unless an addition to be presently described is made to the apparatus. Much more convergence, unless **extra lenses** are added,

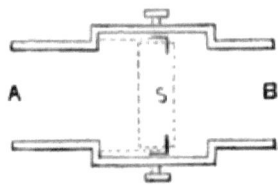

FIG. 181.—Crystal Stage.

causes a great deal of light not to get through the Nicol analyser; and much less gives fewer rings, unless a very thick slice be employed. For a moment's consideration will show that the amount of retardation in a plate of crystal thus cut, depends on both the thickness of the plate and the amount of convergence; and that the rings must become closer and more numerous as the plate increases in thickness. The private student will often find a simple tourmaline pincette (Fig. 182) the most convenient apparatus. A slice of tourmaline is mounted so as to be capable of rotation at each end of the spring tongs, and the crystal plate to be examined is held between them; the rays passing through this simple polariscope into the small pupil of the eye, are sufficiently convergent to exhibit the phenomena. Or a single loose tourmaline, such as the one used in the rotating frame, held close to the eye with the crystal close up to it, will show the rings well, if the whole, and the eye, are turned towards the plane-polarised light from a glass plate or any other polarising surface, or even towards certain portions of a bright sky (see Chapter XVII.).

FIG. 182.—Tourmaline Pincette.

162. **Preparation of Crystals.**—Many crystals can only be prepared, as a rule, by skilled workmen; and the most immense variety, numbering over a hundred, of uni-axials and the bi-axials to be next considered, may be obtained from Dr. Steeg and Reuter, Homburg von der Höhe, who prepare this class of objects for almost the whole world. The rarer crystals can hardly, in fact, be obtained elsewhere; but Messrs. Darker and one or two

Pl. 5.   RINGS AND BRUSHES IN CRYSTALS.

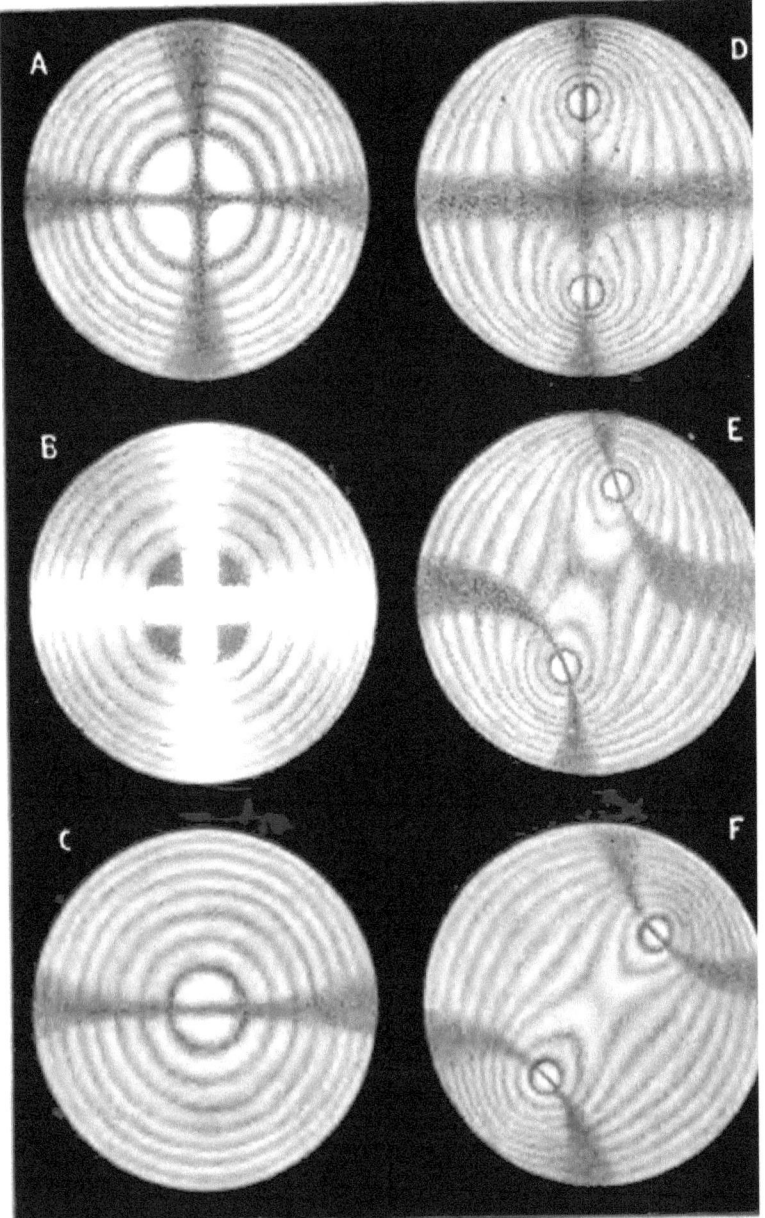

A   Calcite analyser crossed   D   Nitre, showing both axes
B   „   analyser parallel   E   „   slightly rotated
C   Single Axis of Topaz   F   „   axes rotated 45°

other English opticians prepare calcite, quartz, borax, sugar, and some dozen **others of** those most usually inquired for. Several crystals, however, are quite within the reach of **the** amateur. **Nitrate of** soda, if a good clear crystal can be found, is a fine uni-axial. A slice may be **ground on** ground **glass,** polished partially with a little water, and mounted in Canada balsam between two glass circles. Or, in winter, a piece of clear *ice*, $\frac{3}{4}$ inch or so thick, *held* (for it can hardly be placed in the stage) in front of the **bare** nozzle, with the Nicol held in front and rotated, will give beautiful rings. Ferro-cyanide of potassium or prussiate of potash, is a cheap crystal, found in prisms or tablets, and easily splits across the optic axis in slices, which have natural polished faces, and therefore need no other preparation than to be protected **in balsam between two glasses.** It is normally a uni-axial showing circular rings; its other frequent phenomena **will be treated of** presently. Hyposulphate of **potash is another** crystal which gives fair rings, as do phosphates of ammonia or potash. Many of the soft crystals, uni-axial or bi-axial, after grinding on ground glass or stone, only need rubbing once or twice with the wet finger, the balsam in which they are mounted perfecting the polish, as it is nearly the same index of refraction as most of them. Others can generally be polished with a little putty powder or colcothar **on a piece of fine silk.** Sugar must be done dry or with a little oil.

163. **Artificial Crystals.**—Dr. Brewster made *artificial* crystals by melting together equal parts of white wax and rosin, thoroughly mixed, and with a pointed rod dropping two or three drops on a small circle of glass, on which was laid a similar circle, forced down on the composition with a strong pressure. The molecules are thus subjected to strong compression in a direction perpendicular to the plates; and the result is not only double refraction,

but the slide shows rings like a crystal in convergent light.[1]

164. Anomalous Dispersion.—Apophyllite Rings.—From the phenomena of anomalous dispersion (§§ 55, 67) we should expect that some crystals would show exceptional phenomena, similarly due to exceptional retardation of various colours. This is so. Apophyllite, for instance, is remarkable for the fact that it is "positive" for one end of the spectrum, "negative" for the other, and non-doubly refracting for some intermediate colour, generally yellow. Hence we have, instead of rainbow circles as in most cases, rings nearly *white and black*: the usual effect being black rings lined as it were with green only. We have here, therefore, a phenomenon of another kind, due to "anomalous dispersion." By combining slices of certain positive and negative crystals (§ 124), cut of suitable thicknesses, it would be supposed this curious phenomenon should be produced artificially. Such is the fact: but each pair of crystals has to be mutually selected, one with reference to the other, else the counter-action is not sufficiently accurate.

Apophyllite also shows in some specimens a most beautiful tesselated or "mosaic" construction. This, however, is more analogous to the compound crystals mentioned further on. Such crystals have to be specially sought for and selected.

165. Quartz in Convergent Light.—The phenomena of quartz are what we should expect from those already observed in parallel light. Where the light is distinctly convergent, the ordinary doubly-refracting forces come into play; showing circular rings and a cross. Towards the centre, however, the rotatory power of the two circular axial

---

[1] It is very easy to get colour in this way, but in the few trials I have made I have never got perfect rings to please me; probably for want of pressure, as I only used that of my hand.

rays comes more into play; and hence there is never a black centre as in most other crystals, but a coloured area, of a size according to the convergence of the rays, the colours changing with rotation of the analyser. The quartz system is represented in A, B, Plate VII. As a plate of quartz cut parallel to the axis only produces plane vibrations like those of selenite, Sir George Airy suggested that the normal vibrations in such crystals were *elliptical*, of which the circular and plane waves were extreme limits. This theory, when applied mathematically, is found perfectly to account for all the phenomena, including those of superposed plates presently described (§ 178).

166. **Bi-axial Crystals.**—It has **been** stated already (§ 125) that there are many crystals in which neither of the two doubly-refracted rays follows the ordinary law, but both are extraordinary; **the** index of refraction varying **with the** direction of the ray, and the refracted ray not being always in the plane **of incidence.** Such crystals, upon careful experiment, are found to contain *two* directions in which a ray is not doubly-refracted, inclined to each other at some angle; and each of these optic axes is surrounded by a system of rings. Such are accordingly termed "bi-axial" crystals.

The best bi-axial crystals for the ordinary lantern polariscope are those whose axes are not much inclined to each other, so that both systems of rings can be seen **at** once. The four best are nitre, native crystals of carbonate of lead, glauberite, and some varieties of the felspar called adularia. The last must as a rule be purchased; but the only difficulty about nitre is the getting a good crystal to work on. They all look splendid as they come out of the crystallising vats; but as they dry, nearly all spoil by decrepitation and striæ, and *very* few remain clear. A far from perfect one, however, if tolerably clear, will show very fairly. We want a six-sided prism about $\frac{1}{2}$ inch diameter.

Split a slice about a ¼ **inch or more** thick, as square across as possible, with **a knife ; and then** carefully grind it down on a dry stone till about ⅛ inch thick, square to the axis of the prism. **Finish** with a little water on first a roughish and lastly **a. smooth** ground glass, finally wiping off the moisture with the finger. If necessary give it a lick on each side, and again rub with the finger, which will nearly polish it as already described ; then mount it with balsam, or balsam and benzol, between two circles of glass, and finally mount in a wooden slide with some soft cement, that will allow of adjustment square to the axis of the polariscope. When adjusted **with** the line joining the axes parallel to or across the plane of polarisation, and the analyser crossed, **the** appearance is as in D, Plate V. ; with the analyser parallel, **of course the cross is white** and colours complementary. But if with analyser crossed the *crystal* be rotated, the phenomena change beautifully, the cross opening out into hyperbolic curves (E, Plate V.), which, when the axes of the nitre stand at 45° across the plane of polarisation, assume the shape **of** F. The shape of these rings and black brushes can all be mathematically calculated on the principle of interference.

A moment's thought will show that with our usual **moderate convergence** we cannot thus see *both* systems **of rings** in bi-axials whose axes include great angles ; and it is usual to cut such crystals at right **angles to *one* of** the axes, when of course we get approximate circles traversed by *one* arm of the nitre cross, *i.e.* by a straight black brush through the centre. Topaz (C, Plate V.) is a fine crystal cut in this way ; so are borax and sugar.

167. **Apparatus for Exhibiting or Projecting Wide-angled Bi-axials.**—By some addition to our apparatus, however, it is possible to see at once, or to **project,** both systems of rings in wide-angled bi-axials.

Norremberg first invented a system of lenses to accomplish this object, the arrangement mainly consisting of two hemispherical lenses about $\frac{1}{2}$ inch diameter, between which the crystal is placed, backed by other lenses still further to converge the light; and with an additional focusing or projecting " power " on the side next the eye or screen. Fig. 183 is a section of such an arrangement made by Hoffman of Paris, the plate of crystal being shown between the two hemispheres. Such a combination converges the pencil of

FIG. 183.—Convergent Lenses.

rays to a point within $\frac{1}{8}$ inch or less of the plane surface of the first hemisphere, from which point they as widely diverge, but are re-collected by the second similar set, and finally projected through the Nicol analyser by another focusing lens or pair of lenses.

In my own combination, as already stated, each set of converging lenses is a double hemisphere (see Fig. 152); those next the crystal being $\frac{5}{8}$ inch diameter, and each being closely backed by another larger hemisphere. This is about equal to Hoffman's, but not quite equal to the best of the more complex arrangements, in which each set consists of a moderately small hemisphere of very dense glass (much depends upon the density), closely backed up by three plano-convex lenses of gradually increasing diameter, all four of each set being almost in contact.[1]

If the utmost possible angle be desired, a plan must be adopted which we owe to the ingenuity of Professor W. G. Adams. It consists essentially in forming the two small hemispheres into the front and back sides of a *cell*, which is filled with oil in which the crystal is immersed. The oil of course nearly prevents the much greater divergence of the oblique rays caused by refraction from or into air, the direction in the oil being little altered. In this way the extreme angle of 90° can be collected, but the method is only suitable for private experiment. Without the oil, extreme angles in highly refractive bodies would be kept within the crystal by total reflection.

The rings and brushes can be finely shown in any microscope which possesses a draw-tube and the customary wide collar under the stage. The usual arrangement has been an addition to the eye-piece; but this is very imperfect in performance, only sufficing for small angles. Messrs. Swift and Son have very lately contrived a far superior arrangement. It consists of a small hemisphere, whose plane side (turned upwards) is brought flush with the stage, and has below it another strongly convergent lens and the polarising

---

[1] Without pronouncing upon the general merits of any optician, I may state that the very best actual arrangement I have ever seen was constructed by Laurent, of Paris. It not only embraced unusually wide angles, but was remarkable for flatness and equal illumination of field, perfect freedom from colour, and sharpness of the rings. From it I take the description of the "best" arrangement as above; but the arrangement of lenses was in the first instance due to Descloizeaux.

Nicol. This forms the converging arrangement, assisted if necessary by mirror or condenser; and the objective answers for the second set, with the addition of **another** lens in the end **of** the draw-tube, **in** which also a quarter-wave plate can easily be inserted. The cost of this addition is only about 35s., and its performance is excellent, as it will pick up with rather a short focus angles of nearly 45°, or with extreme care even a little more. All the range of phenomena in this and the following chapter are therefore brought within **each** of the microscopist at very little extra cost.

For a convergent apparatus like Fig. 152 or Fig. 183 very small crystals suffice, the light being converged to a mere point. Some of the best crystals can in fact only be obtained large enough to give sections about $\frac{1}{8}$ inch across; but such sizes are perfectly shown if blacked round.

168. **Angles of Crystals.**—The following is a list of a few good crystals most easily procured, the angles being quoted from various authorities. These are the real angles of the axes in the crystal itself, but the *apparent* angle is much increased by the refraction from the perpendicular into air of the oblique rays. The angles vary somewhat, (§ 177), and scarcely two authorities quote the same.

Glauberite	2° to 10°
Lead Carbonate, or Cerussite .	8°·07′
Nitre	7°·12′
Arragonite	18°·18′
Titanite	30°
Borax	38°·32′
Mica [1]	45°
Sulphate of Zinc . . .	44°·4′
Topaz	50°

[1] **Mica** differs exceedingly, from a uni-axial up to as much as 75° (see § 177).

Sugar	50°
Selenite	57°·30′
Nitrate of Silver	62°·16′

169. **Anomalous Dispersion in Bi-axials.**—The phenomena analogous to anomalous dispersion are still more remarkable and interesting in bi-axial crystals than in uni-axials. Not only are the axes as a rule somewhat differently inclined for different colours, but in some bi-axial crystals they do not even lie in the same plane. Borax is a good case of this kind. If a plate cut across both axes be placed in the polariscope, with the line joining its axes parallel with or perpendicular to the polariser, when the analyser is crossed it will be found impossible to produce a *perfectly* straight black brush in the line of the axes; both arms are perceptibly curved, and when the long arm is perpendicular there is a perceptible tint on the left of the top arm, corresponding to one on the right-hand of the bottom arm. In Adularia the centres of the "eyes" for red and blue are dispersed on lines parallel to each other, and perpendicular to the long arm of the cross. In other crystals, whose axes lie in the same plane for all colours, the inclination varies very greatly, and progresses in reverse order for some crystals compared with others. The effect of this is, that when the plate is turned round so that its axial images are at 45° angle with the polarising plane (Fig. F, Plate V.), the parabolic "brushes" are in nitre margined with red inside and blue outside, while in carbonate of lead the reverse is the case. In carbonate of lead, and many other crystals, the dispersion of the axes is so strong that the hyperbolic brushes appear only as red and blue, there being no black brush whatever; while the rings in white light assume very peculiar forms. In monochromatic light, however, true lemniscates will be observed.

But the most remarkable example of anomalous dispersion in the axes is in a few crystals such as brookite, in which the axes for red light lie in one arm of the cross, while those for violet lie in the other arm; or, in other words, the axes for the two ends of the spectrum lie in rectangular planes! Hence the figure presented in white light somewhat resembles that of a "crossed" crystal (§ 176). Brookite is not at all easy to procure of sufficient size; and is extremely difficult to project owing to its red and somewhat opaque colour; but similar phenomena may be seen in the *sel de seignette*, or triple tartrate of soda, **potass,** and ammonia, which is colourless and transparent. Viewing or projecting this crystal, with even a pale red and blue glass alternately in the large slide-stage of the polariscope, the rectangular planes of axial dispersion will be readily seen; and if a cobalt glass can be procured of the right shade and thickness to cut out the middle of the spectrum and leave the blue and extreme red only, the two systems of rings can be seen superposed by a single observer, though the light is insufficient for projection. In white light the figure is a very beautiful one, partially resembling that of two ordinary bi-axials crossed, but with modifications very difficult to describe.

170. **Theory of Bi-axials.**—The theory of bi-axial crystals was gradually elucidated by the labours of Brewster, until Fresnel framed the one simple general conception of three *axes of elasticity* in three rectangular directions. If all these were equal, rays proceeding from a point in the crystal in all directions would at the same moment reach the surface of a sphere, and the crystal could have no double refraction. If two were equal, both being perpendicular to the third, *all* perpendicular to that third must also be equal, and the crystal must be uni-axial. If all three were unequal, it was shown by a beautiful mathematical

analysis that the crystal **must be** bi-axial, the axes of **no** double refraction being simply resultants **of the** three different elasticities, and lying in the plane of the greatest and least elasticities, **at an** angle **dependent on the** relative magnitude of **the** third or mean elasticity. This beautiful theory was shown to be exactly what ought to follow **from** the simple assumptions of the undulatory theory, and to account for every detail of the known phenomena; but several necessary deductions followed, which did not become apparent till after Fresnel's death. It is therefore interest**ing** to see how far these purely theoretical deductions were justified.

171. **Mitscherlich's Experiment.**—It followed, first of all, that any alteration of the respective elasticities must necessarily alter the inclination of the axes; and that should the change go so far that the mean elasticity became first **equal** to either of **the others, and** then reversed its relative magnitude **compared with it,** the crystal must first become uni-axial, **and afterwards bi-axial** with axes in a plane at right angles to the **first.** Now we have already seen that heat will effect such changes; selenite especially **being considerably** altered by heat in relative dimen**sions and** relative elasticities, **by** a very moderate rise of **temperature.** Professor Mitscherlich therefore took a plate **of this** crystal cut so as to show both the axes; and ex**posing it** gradually to a heat not exceeding that of boiling water, he **had** the satisfaction of seeing the **two** systems of **rings** gradually approach each other. A point **was soon** reached at which they coincided, and *the crystal became uni-axial;* and the moment after the axes began **to** open out in a direction at right angles to **the former.**

This fine experiment is projected with the greatest ease, the only difficulty being in cutting the crystal. Unlike **mica, whose** natural laminæ **are at right** angles to the two axes, the laminæ of selenite include them, and the crystal has therefore to be cut across its cleavages. A plate of copper or brass, A (Fig. 184), has a hollow turned in its centre so as to leave only a very thin plate of metal to support the crystal, an aperture sufficiently large for the rays being bored in the thin plate. The crystal, rather larger than the aperture, should be set in a disc of cork rather thicker than itself (shaded black in the figure) and shaped to fit the hollow; the whole being covered by another thin metal plate, C. The arrangement is then placed on the stage of the projecting apparatus, so that the ends of the metal project well, and these ends are heated by spirit-lamps, or by one lamp alternately.[1] Suppose the crystal is so arranged on the stage that the two systems of rings

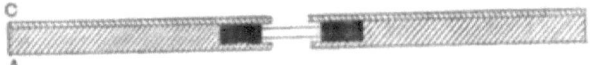

FIG. 184.—Slide for Mitscherlich's Experiment.

appear at first perpendicularly upon the screen. As the slide warms there will probably be a cloudiness for a minute or two, owing partly to moisture on the lenses and partly to heating of the air between the laminæ of the crystal; and the heat should be applied gently till this clears away, as it will soon do. Then the axes will gradually be seen to approach, till the rings exactly resemble those of a plate of calcite;[2] and after this they open out again in a *horizontal*

[1] Another good way is to cut a hole, large enough to hold the crystal loosely, quite through the middle of a slip of brass the same thickness, retaining the crystal by doubling over both sides a piece of card in which are cut holes too small for the crystal to fall through. The ends of the brass are better bent away from the stage. Crystals cost 3s. to 5s. each.

[2] At this stage the lamps had better be removed till it is seen how things go. The metal slide will probably have taken up sufficient heat to go on further without more; and a very moderate excess of heat will calcine the clear crystal into mere plaster of Paris.

Y 2

direction. On the lamp being removed the whole process is reversed.

172. Conical Refraction.—A more surprising proof of the reality of these "wave-shells" was discovered by Sir W. Hamilton. On projecting them according to this theory, they were found to resemble those partially shown in Fig. 185 (from Müller-Pouillet). Now if a single ray traverses the crystal in the line P P, or P′ P′, on reaching the surface it is refracted into air, as usual, from the perpendicular. That perpendicular has reference to the *tangents of two different*

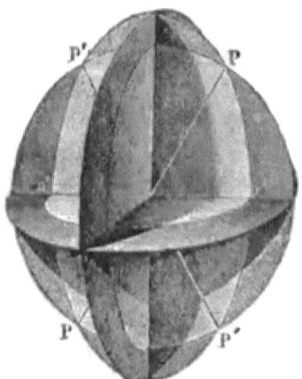

FIG. 185.—Bi-axial Wave-Shells.

curves, and so produces two different refractions. If the shells are completed, however, as in a solid model, it is found that the four points P are *cusps*, or hollows resembling that surrounding the stem of an apple;[1] and it therefore follows that on emergence from the point P, the ray must be spread out into a *diverging conical shell of rays*. Here, therefore, was a mathematical prediction of a phenomenon

---

[1] It is very difficult to give a clear idea of the complicated wave-surfaces in bi-axials to ordinary readers. Some may derive assistance from another and differently shaded figure, which will be found under the article "Undulatory Theory," in *Chambers's Cyclopædia*.

never foreseen by Fresnel, who had confined himself **to the** single plane through the points P **shown** in the diagram; and such a kind of refraction was not only opposed to all experience, but to all apparent probability.

At Sir William Hamilton's request the matter was tested by Dr. Humphrey Lloyd in a crystal of arragonite. **The** lines P P and P′ P′ are lines of single velocity, and coincide *nearly* with the optic axes, but not exactly, their angle being nearly 20°. A plate is needed ¼ inch or a little more in thickness for this experiment, cut across both **axes.** Then a thin metal plate with a very small **aperture was fixed on**

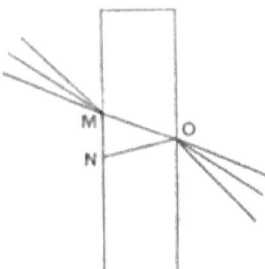

FIG. 186.—Conical Refraction.

one side of the crystal plate, and a movable one on the other, while the crystal was fixed in a frame movable by a screw so as to present **the** crystal at various angles. A beam of light was then condensed on the aperture O (Fig. 186) at an angle of 15° or **16°,** so **as** to be refracted in the direction of O M, or O **N, which** is nearly an optic axis.[1] When the adjustment was complete, on looking through the second aperture, instead of two apertures a *brilliant ring*

---

[1] The *convergence* of the rays, to produce a single ray within the crystal, must of course correspond with the calculated *divergence* on leaving the crystal. The optic axis is the normal to the common tangent plane touching both wave-surfaces. (See next page and § 125.)

of light appeared to the observer; Sir William Hamilton's prediction being thus justified to the letter!

In a similar way it was shown that if a very small pencil of parallel rays (sensibly apparent as a single ray) were incident at o, so as to be refracted exactly into the optic axis, it should be divided into an *internal* cone within the crystal, which on emergence would resume parallelism and appear as a *hollow cylinder* of rays, since the cone would reach the wave-shells at the points where the common tangent-plane touched them in a small circle surrounding the cusp. The cone in this case was so small that the phenomena were much more difficult to observe: but by adjusting the crystal with extreme care, this prediction also was verified in actual fact.

The easiest way of observing conical refraction is to substitute for the aperture o, a *fine slit*, in the plane of the optic axes. Then if the aperture M be fixed on the other face tolerably near the correct position, on gradually tilting the crystal in the plane of the axes, when the right point is reached the slit will be seen to split in the middle into two oval curves. The experiment is only one for the student, the small pencil of light not permitting of projection.

173. **Relation of the Axes in Uni-axial and Bi-axial Crystals.**—It further follows that, according to this theory, the optic axis of a uni-axial crystal by no means coincides in character with *one* of the axes of a bi-axial; but is simply a limiting case, in which *both* these axes coincide. Professor Mitscherlich's experiment is one beautiful proof that this is the case: and further optical proof of it will be found in the next chapter, where it will be seen that important optical phenomena of both axes in a bi-axial, are preserved distinctly in uni-axial crystals.

174. **Composite, Irregular, and Hemitrope Crystals.**—Twin or macled crystals are found very often: a slice of nitre, for instance, can be found without difficulty

that will exhibit four systems of rings. Calcite is often found in which thin layers *crystallised in the opposite direction* are frequent. A large crystal of such calcite, if a ray is sent through it to the screen, gives a greater or less number of coloured images: the interrupting film answering to the films of mica or selenite in our earlier experiments, and the thicker masses to analyser and polariser—the whole being a sort of natural Huyghens' apparatus fixed in one position, with a film between the two prisms. But far more beautiful effects are obtained if a plate be cut across the axis including one of these films or planes, and examined in the convergent light. The system of rings is then modified by glorious brushes of coloured light, which radiate somewhat like the spokes of a wheel (D, Plate I.); and the pattern of the rings themselves may be varied in a beautiful manner. If such a plate is cut thin enough to exhibit in the very convergent system, on moving it about so that the conical pencil may traverse different points in succession, the phenomena will sometimes vary in an extraordinary degree.

These effects may be partially imitated by placing a film of mica or selenite between two thin plates of calcite cut across the axis; and are perfectly imitated if the calcites are ground into a pair of wedges making together a parallel plate. Pieces of calcite subjected to strong pressure and then cut, also give fine irregular phenomena; or a plate may be projected whilst squeezed in a small vice.

Quartz crystals are often found in which right and left-handed crystallisation are combined. In amethyst they are often combined in alternate layers. Such a crystal shows of course bands or other portions which change in colour in opposite directions. These are best shown in parallel light.

Other irregularities are often found. Ferro-cyanide of potassium, for instance, is properly a uni-axial crystal. But crystals can be found, out of any jar, of which slices give bi-axial effects: and yet others, which **give** symmetrical coloured patterns, which are very beautiful though crystallographically irregular.

175. **Mica and Selenite Combinations.**—Norremberg demonstrated the cause of these beautiful phenomena, or **at** least closely imitated them, by combining films **of** selenite with films of mica, a selenite being placed between **two micas.** The micas in Norremberg's arrangements were **parallel** ; and in some combinations the principal axis in the selenite was parallel to the mica axis, and **in** some across it. Then four of these triple films were combined in any way, so that the two kinds were *symmetrically* superposed : such as four of each, two of one and two of the **other, two of each** crossed, two pairs **of the same** crossed, one of **each** crossed alternately, and so on. The student will probably content himself by crossing films **experimentally** ; only taking **care that pairs of the mica are of** equal thickness. By trial he can hardly fail to **find** positions which give beautiful effects; and the preparation can then **be** permanently mounted in balsam and benzol. The micas must always be either parallel, **or** exactly crossed ; and Mr. Fox, who has experimented very largely in this direction, tells me that little effect can be got with less than about $\frac{1}{2}$ wave thick, or thicker, unless many films are employed. Bi-axial constructions, made with films whose mica-axes are all **the same way, are** still more resplendent in colour, but less **attractive** in figure, than the crossed patterns; a poor idea **of two of** which is given in **E, F,** Plate **I.**

176. **Crossed Crystals.**—These give beautiful effects, but the best of them require the convergent system. Two plates of mica are easiest put together, and give four systems of rings (A), or when the preparation is rotated **45** degrees show beautiful hyperbolic curves in the centre **(B,** Plate VI.). Crossed arragonite or topaz give similar

Pl 6. CROSSED AND SUPERPOSED CRYSTALS.

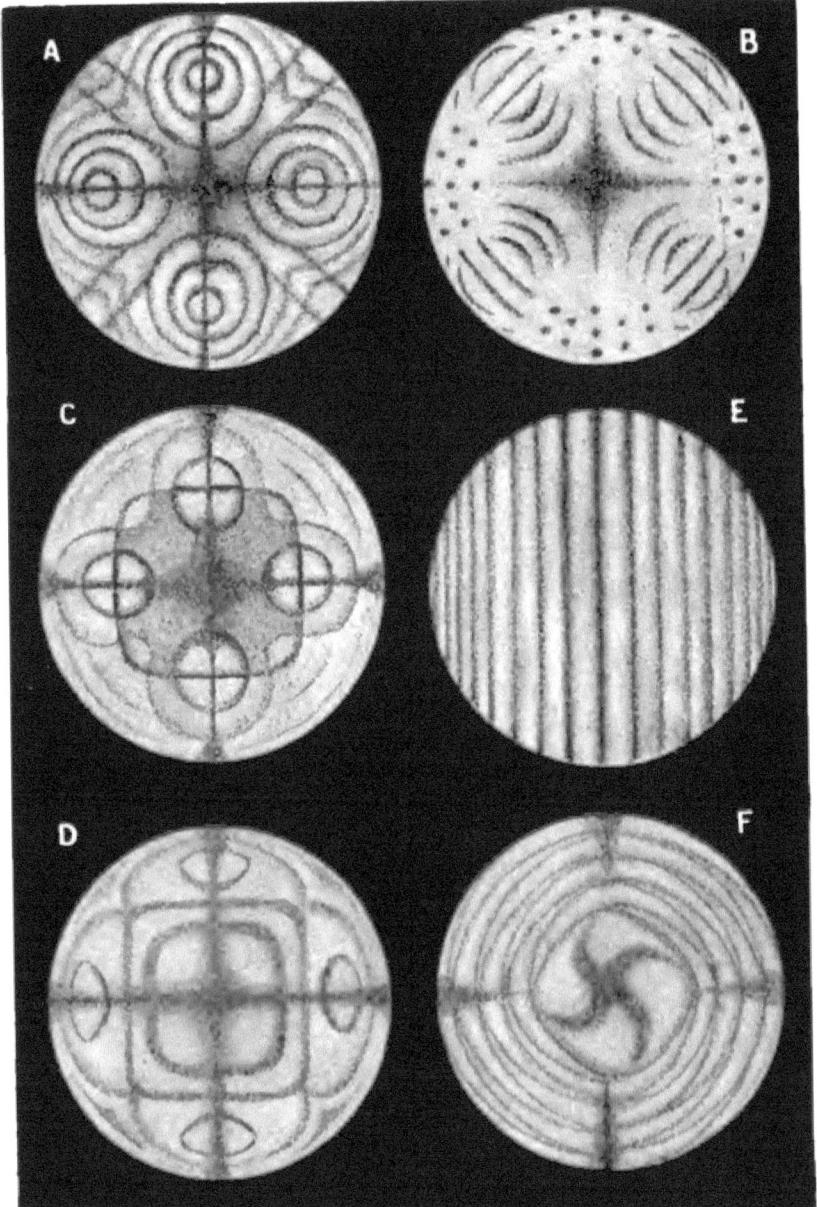

A Two Micas crossed
B Ditto axes rotated 45°
C Four Micas crossed
D Eight Micas crossed
E Savart's Bands
F Airy's Spirals

effects, without the smokiness of the mica. Crossed titanite is effective, owing to its peculiar dispersion.

Two ordinary films of selenite 1 mm. or more thick, too thick to show any colours alone, when crossed give hyperbolic curves in convergent light; as do two quartzes parallel to the axis.

177. **Norremberg's Uni-axial Mica Combinations.**—The most interesting, however, of this kind of combinations are also due to Professor Norremberg. Mica is found occasionally *uni-axial;* and bi-axial specimens are found of all angles from 0° up to 75°, in this respect resembling the ferro-cyanide of potassium. Senarmont had proved by experiment the results of combining salts crystallising in identical forms geometrically opposite. Thus, the double tartrate of soda and potash (Rochelle salt) crystallises in prisms; and if we replace the potash by ammonia, we get similar prisms. Moreover, both are bi-axial crystals with an angle of 76°, and a peculiar dispersion of the axes pointed out by Sir John Herschel is the same in each. The one optical difference is, that the plane of the optic axes passes through the smaller diagonal of the rhomb which forms the base of the prism in one case, and of the longer diagonal in the other. Senarmont showed, that by crystallising mixtures of the two double salts, the angle of the axes could be diminished; and that with a certain proportion the crystal became uni-axial like calcite; though, owing to the great dispersion of the axes, it can only be strictly uni-axial for one colour in any given combination, whence the peculiar dispersion noticed in 169.

Hence Norremberg supposed there might be two kinds of micas; isomorphous, but geometrically opposite: and that the variable angles, and uni-axial micas, might be produced by superpositions of infinitely thin films of each, in different proportions and positions.

Experiment proved this. A number of thin films of mica crossed alternately at an angle of 60°, reduced the angle of the bi-axial mica from about 70° to about 46°. But far more interesting was the gradual passage to the uni-axial form when the films were crossed at 90°. Denoting a thickness of one wave-length retardation (*i.e.* four times that of a quarter-wave plate) by w, Norremberg constructed the following series, where the figure under the alphabetical letter denotes the number of films, and the bottom fractions the thickness, the total of all being three wave-lengths of retardation.

A	B	C	D	E	F
1	2	4	8	12	24
$3w$	$\dfrac{3w}{2}$	$\dfrac{3w}{4}$	$\dfrac{3w}{8}$	$\dfrac{3w}{4}$	$\dfrac{3w}{8}$

In this series of preparations (which require the converging system of lenses to exhibit them), A of course gives the ordinary bi-axial phenomena. B gives four systems of rings with hyperbolic curves (A, B, Plate VI.). In C we get the first approach to the uni-axial character, which becomes more and more perfect as we proceed, until at last there is absolutely no difference between the phenomena and that of a plate of calcite. It will also be found that C and D give remarkable resemblances to the irregular crystals of the ferro-cyanide (A, C, D, Plate VI.).

There is no magic in three wave-lengths as the total thickness. All that is necessary in making these instructive preparations is that, as before described, all the films of any one preparation be of the same thickness, and the planes of polarisation which contain the two optic axes, in each alternate film, accurately crossed. The method of obtaining both conditions has been already given (§ 157). I prefer to add an intermediate preparation of six films between

C and D, which gives the first complete black square ring. Then eight films give **two** such rings, the inside one more circular and **the** outer one square; while twelve films give three rings, sixteen four rings, and so on.

178. **Airy's Spirals.**—Owing to the character of the doubly-refracted rays in quartz, if a plate of "right" and another of "left," of equal thickness, be placed together in the crystal stage and analysed by convergent light, the colour is not exactly neutralised, as it is in parallel light, but we get a most beautiful quadruple spiral (shown in F, Plate VI.). These spirals were discovered by Mr. Airy, and are called by his name. They form a most beautiful screen projection. About $\frac{1}{16}$th **of** an inch to $\frac{1}{8}$ inch each is a good thickness for each plate for the objective described, the plates of all crystals needing to be thicker the **longer** the focus employed. A *single* **quartz** plate will, however, also show **the spirals** in Norremberg's "doubler." If truly cut square to **the axis,** it may be laid **on the bottom** mirror, holding a lens about $1\frac{1}{2}$ inches diameter **and, say, 2 inches** focus above it, **so** as to converge **and** re-collect **the rays** which have twice passed through the plate; but should **the** quartz not be quite true, the lens must be laid *on* it, and both together adjusted by hand till accurate spirals appear. This experiment is particularly interesting as showing the reversal of the rotation described in § **150.**

Reusch's artificial mica-film quartzes present **all** these phenomena *perfectly*.[1] **In** convergent rays **a** single one exhibits the rings with coloured centre, and black brushes as the centre is receded from; and **a** pair of right and left superposed give perfect Airy's spirals.

179. **Savart's Bands.**—If slices of quartz **are** cut neither parallel with axis nor across it, but at 45° angle,

---

[1] I have specimens, prepared by Mr. Fox, which **are** as perfect in rings, and brushes, and Airy's spirals, as any quartzes could possibly be.

such slices singly show hyperbolas; but if two such slices are crossed at right angles, these become in convergent light straight lines, called "Savart's bands," and such a combination is a very sensitive test of polarised light, the least degree of polarisation making the bands visible. Slices of calc spar split naturally and crossed give the same results (E, Plate VI.).

All these phenomena of quartz have been worked out mathematically from the theory, most of them by Mr. Airy.

It may be mentioned here, that in these phenomena, as in all others, if the thin glass analyser, Fig. 155, be used instead of the Nicol, the complementary image to that on the screen will always appear on the ceiling. Complicated as the phenomena may be, such an analyser never fails thus to sort them out rigidly into two complementary sets.

ROTARY AND CIRCULAR POLARIZATION

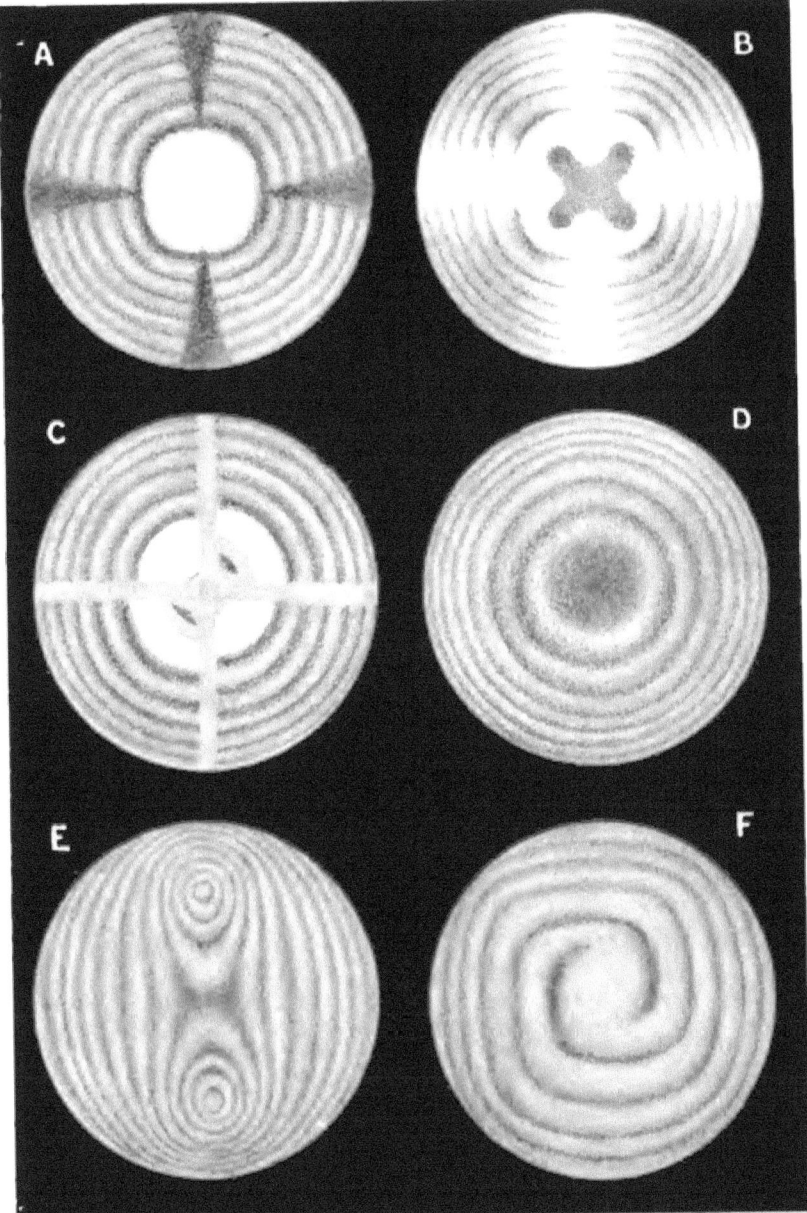

A Quartz with analyser crossed   D Calcite, polarised & analysed circularly
B   analyser parallel   E Vitre   Ditto   Ditto
C Calcite, polarised circularly   F Quartz in convergent circularly polarised ligt

CHAPTER XVI.

OPTICAL PHENOMENA OF CRYSTALS IN CIRCULARLY-POLARISED LIGHT.

Modifications in Crystal Figures Produced by one Quarter-Wave Plate—Explanation of the Phenomena—Results of Polarising and Analysing Circularly—Quartz in Circularly-Polarised Light—Spiral Figures showing the Relation of Uni-axial and Bi-axial Axes.

HAVING traced the general phenomena of crystals when examined in convergent plane-polarised light, we next proceed to examine the appearances they present in circularly-polarised light; being prepared by our previous experiments with this description of light for some interesting variations of the phenomena.

180. **Effects of a Single Quarter-Wave Plate on the Rings of Crystals.**—We place first the large quarter-wave plate (in its usual position) in the ordinary slide-stage of the polariscope; and place the plate of calcite cut across the axis in the crystal stage. The calcite loses its black cross, as we should expect, the cross being replaced by a thin (mere lines) nebulous *grey* cross which rotates with the analyser, on either side of the arms of which alternate quadrants of rings are *dislocated* as in C, Plate VII., the light rings in one quadrant being opposite the dark parts in its neighbours. This figure does not change in the least as the

analyser is rotated, but the quadrants and rectangular nebulous lines simply rotate with it.

Bi-axial crystals give similar **phenomena**, the arms of the cross which pass through the axes being replaced by nebulous lines, on each side of which the *semicircles* of each system of rings are dislocated. This is most distinctly shown by placing in the crystal stage a plate cut across one axis only, as in the nearly circular rings of one axis in sugar-candy. If the quarter-wave plate be rotated 90°, the quadrants or semicircles which were first the smallest, will gradually become the largest, and *vice versâ*. Precisely similar effects will be found if the large quarter-wave plate be withdrawn, and the light *analysed* through the smaller one, placed in its usual position between the crystal and the analyser.

181. **Explanation of the Phenomena.**—This strange dislocation of the rings is easily explained, if we remember the original composition of the circularly-polarised ray, which is compounded (in passing through the mica-film) of two rectangular plane vibrations, one of which is a quarter-undulation in phase behind the other—or in other words, referring to Fig. 171, one vibration in the middle of its swing acting upon another at the moment of rest, represented by the arrows in that figure. In the quarter-wave plate, supposing the plane of polarisation to be vertical, the vibrations are diagonal; and may be represented by the diagonal pairs of arrows in Fig. 187. Let this figure represent the plate of calcite with polariser and analyser crossed: in each quadrant the circularly-polarised ray, on entering the crystal, is doubly refracted into its plane-polarised components, one of which *enters* the plate a quarter-undulation behind the other. But we have already seen (Fig. 180, § 160) that the crystal itself doubly-refracts any non-central ray into two rectangular components represented by radii and tangents to the circles; and in every uni-axial crystal, either the radii or the circles are

uniformly the most retarded—in **the** case of calcite the
radii. Now in each order of colours or "ring," there is some
position or distance from the centre where the retardation is
either a half-wave or an **odd** multiple **of a half-wave;** and
on the other hand, mere inspection of Fig. 187 shows that
in the middle of each quadrant the two plane **vibrations**
emerging from the mica-film (and after composition into a
circular orbit, again decomposed) coincide with the plane
vibrations—radius and tangent—in the plate of calcite. It is
also seen that whereas in the calcite alone, all the radii are

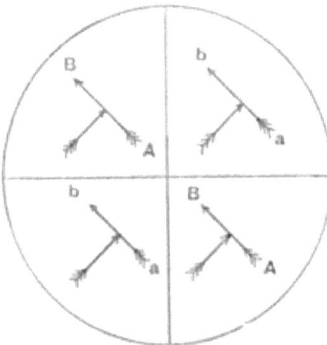

FIG. 187.—Effect of Quarter-Wave Plate.

at given intervals retarded one **or** more half-waves behind
their tangents, it is different with **the mica** retardations. In
two quadrants the vibrations, **A B, A** B, coincide with radii,
while in the alternate ones, *a b, a b,* coincide with tangents.
All the *uniform* annular half-wave retardations of the calcite
are therefore in two quadrants augmented by an additional
quarter-wave retardation from the mica; and in the intermediate quadrants *counteracted* by a quarter-wave retardation
of the opposite polarising plane. The result is therefore
a dislocation by two quarters, or half a wave; which
brings the bright rings of one against the dark rings of its
neighbouring quarter.

182. **Result of Polarising and Analysing Circularly.**—This having been demonstrated, place *both* quarter-wave plates in their usual positions, and the calcite in the crystal stage between them. Another very beautiful modification follows. With analyser crossed or parallel, all cross lines or dislocations have vanished, leaving only perfect *circular* coloured rings (D, Plate VII.). But as the analyser is rotated, alternate quadrants *expand* and *contract* in a beautiful manner, till at 45° we have the dislocated quadrants, and at 90° the *complementary* perfect rings to those in the

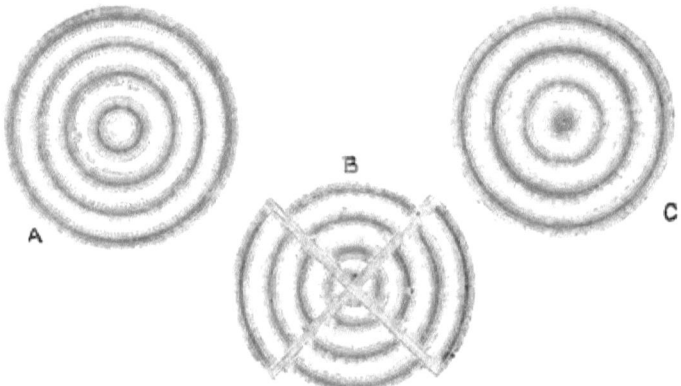

FIG. 188. Uni-axial Crystal Circularly Polarised.

first position; the successive appearances being represented in Fig. 188, A, B, C.

The appearances in a bi-axial, such as nitre, are analogous and beautiful, all brushes having vanished, and each axis being surrounded by unbroken rings (E, Plate VII.). And lastly, at the point when the rings are perfect and unbroken, if either the crystal stage with all it bears—crystal, quarter-wave plate, and analyser—or even only the quarter-wave plate and analyser, be rotated, no change whatever occurs; the rings *remain* unbroken, and all trace of polarising planes, as theory

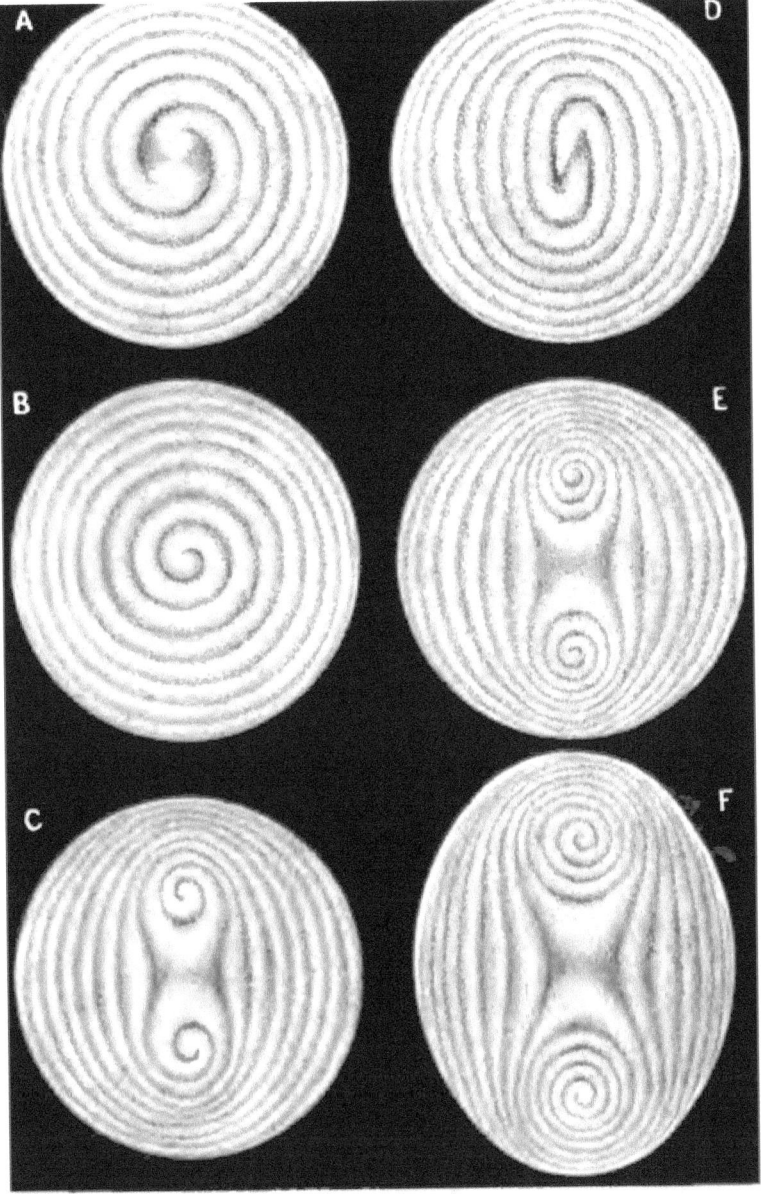

A Calcite
B Single Axis of Sugar
C Nitre
D Thin Nitre in highly convergent light
E Arragonite,   Ditto
F Mica of 70°   Ditto

would lead us to expect, is completely lost. The most marked demonstration of this is to rotate the whole affair with a crystal of nitre, **or** arragonite; **as** nothing can more strikingly show the perfect independence of any planes of polarisation than the fact that such a bi-axial, which ordinarily changes so greatly (see D, E, F, Plate **V.**), **when** the crystal itself is rotated till its two axes stand at 45°, shows no change whatever when rotated as described. **The** perfect rings round the two axes rotate round each other, but remain in all positions unaltered.

It may be worth while to remark, that a circular chilled glass in the optical stage behaves exactly like a uni-axial crystal in the crystal stage, under circularly-polarised light. Placing next the polariser a quarter-wave plate, and then the chilled glass, the usual black **cross is** replaced by the thin nebulous grey **cross, which** rotates unaltered with the analyser; and adding the second quarter-wave plate, the analyser at 0° and 90° gives perfectly unbroken and complementary rings, while **in** intermediate positions the alternate quadrants **expand and** contract during **the** transition.

183. **Quartz in Circularly-Polarised Light.**—When quartz is placed in the crystal stage and examined **with a** single quarter-wave plate, the curious double spiral shown in F, Plate VII., is seen. **The** number of spiral rings depends **on** the thickness of the plate and convergence of the rays. With two quarter-wave plates—one each side of the quartz—we get perfect rings, as with other uni-axial crystals. **The** significance of this phenomenon will presently appear.

184. **Spiral Figures.**—It is remarkable that in the foregoing experiments the rings surrounding a single axis of a bi-axial crystal, when polarised and analysed circularly, appear as perfect circles; or are apparently precisely similar to those

z

surrounding the axis of a uni-axial crystal. From this the conclusion might be drawn that one axis was similar to the other in character. We have seen already (§ 173) that this could not be the case according to Fresnel's theory; and Mitscherlich's experiment (§ 171) is a beautiful demonstration of the gradual approach of the axes of a bi-axial until the uni-axial form appears as a limiting case, in which both axes coincide. It appeared to me, however, desirable to seek for further experimental proof that the axis of a uni-axial crystal was not only such a limiting case, but actually did still retain or embrace within itself in some visible form the optical characteristics of the two axes thus brought into coincidence. This object seemed most likely to be obtained by the aid of quartz or some similar substance having properties of rotatory polarisation. Such substances having, apart and distinct from the ordinary double-refraction, two different *axial* velocities or waves, capable of being brought into interference; and the two axes of a bi-axial being dissimilar from the nature of the case, one of them having the character of a principal axis and the other being secondary to it; it seemed probable that, by proper means, the two axes might be made to exert some kind of differential or selective action upon the two sets of waves passing through the rotatory substance. This expectation was confirmed by the curious double-spiral (first noticed by Mr. Airy [1]) as displayed by quartz when placed in somewhat convergent circularly-polarised light (§ 183), the true cause of which appeared to me to be connected with this very matter, as we shall presently see that it is.

I placed therefore in the apparatus first, next to the polariser, a quarter-wave plate; then, in the crystal stage, a calcite; and next to this, also in the crystal stage (made with a wide opening to admit extra plates) a plate of quartz

[1] *Cambridge Transactions*, 1831.

from 5 mm. to 7½ mm. thick. The ·result is the beautiful system of double spirals, mutually enwrapping each other, shown in A, Plate VIII., and which changes in colour, or appears to approach or recede from the centre, as the analyser is rotated; though there are of course only certain opposite positions of the polariser (related to that of the quarter-wave plate) which produce them. The point to be here remarked is the *double* spiral observed in the **uni-axial** crystal.[1]

This figure however so closely resembles, in all but the number of its convolutions, that observed by Mr. Airy in quartz alone, in convergent light, that it might possibly be due to the quartz itself. To determine this point it is necessary to test other crystals, and to get rid of any such possible cause. We therefore replace the calcite in the crystal stage by a plate of sugar cut across *one* of its two axes.[2] The result is the spiral shown in B, Plate VIII. It will be observed that with this single axis of a bi-axial we no longer have the double spiral, but a *single* one, corresponding strictly to the theoretic relation of which we are seeking proof, and also showing that these figures are not proper to the quartz in itself, but to some selective action of our other crystals upon its two axial waves and their interferences.

A single axis being thus tested, we replace the sugar in the crystal stage by some bi-axial, such as nitre, cut at right angles to the median line between *both* axes. The result is

[1] These spiral figures were first publicly exhibited at a meeting of the Physical Society of London on November 12, 1881, having been discovered and projected before a few friends two years previously.

[2] It is very difficult to cut sugar in this way, the crystal being so fragile, and the cut quite an artificial one. But the crystal is peculiarly suitable for the purpose, having very little *axial dispersion*, so that one of its axes, if truly cut, gives sensible circles. Sugar is easily cut to show both axes, such a section being parallel to a natural face.

the double spiral shown in C, **Plate VIII**. The relation still holds good; each axis has its own spiral, and the two now mutually enwrap each other as in the calcite.

To examine crystals with wider angles, **we must** of course employ the convergent system. But a moment's reflection will show that we must also alter our other arrangements. In such strongly-convergent light the rings and spirals proper to the *quartz itself*, which have not appeared in the moderate convergence so far employed, would overpower and distort those belonging to the crystals under examination (a single experiment will show that they do). It is moreover well to demonstrate absolutely that the figures are in no way **due to any** effects of *convergent* rays traversing the quartz, but solely **to selective action** upon the right-handed and left-handed **waves** traversing it axially. We therefore **reverse our** arrangements, placing **a** large **plate** of quartz, say $7\frac{1}{2}$ mm. thick, **next** the polarising **Nicol, in parallel** light, and **introducing the quarter-wave plate between the crystal to be examined and the analyser. Of course, as it is the analyser which is now related to the quarter-wave plate, the spirals only appear in opposite positions; while on the other hand, with analyser in those positions, they only move and change in colour as the *polariser* is rotated.[1]

First we take again the small angle of** a nitre crystal, but **which, for** this strongly convergent light, to show conspicuous rings, **must be cut much thinner than the former.** The result **is shown in D, Plate VIII.; where it is to be** observed that we **can** hardly distinguish **its spirals** from those of **the** calcite (A, **same** plate). They **are just** a little drawn **out into an oval** form, precisely **as we should expect**; and that is all. Arragonite, with a **wider axial angle of** about $18\frac{1}{2}$

---

[1] This arrangement might **have been** adopted all **along;** but the experiments are given as first made, **in order to** show how each point was successively determined.

degrees, shows a spiral of several turns round each **axis** (F, Plate VIII.), but still mutually enwrapping each other **at** last; and finally mica, or sugar, or other crystal with **an** angle of say $45°$ to $60°$, shows still **more numerous con**-volutions, **but** still preserving the same mutual relation **(F, Plate** VIII.). Only crystals which, owing to peculiarities in their double refraction for various colours (§ 169), do not show in the ordinary manner tolerably complete lemniscates, for the same reason fail to show these figures.

That the spirals **are** due to selective **action** upon the coloured components of the two axial quartz waves, is shown by the fact that they are not seen at all **in** homogeneous or one-coloured light.

We next examine a crystal **of selenite** gradually **heated,** thus repeating Mitscherlich's beautiful **experiment with this** additional **method of analysis.** Care is required to produce equally perfect figures, the least excess of heat on any side of the crystal **of** course causing distortion. This can however be avoided. **We** first obtain two spirals exactly similar to those of **F,** Plate **VIII.** As the axes **approach and** coincide, the spirals also approach in their centres, until at the point of coincidence they exactly resemble those of the calcite (A, Plate VIII.). And finally they re-open in a direction at right angles to the former. All through we have a *double* spiral; and we can only get a single one by taking separately *one* of the axes **of** a bi-axial; the axis of a uni-axial always preserving what **we** may call its twin character. Thus we have **the** ocular proof sought, of the relation predicated by the theory of Fresnel between the axes of the two classes of crystals.

But the reason is also thus demonstrated of the spirals observed by Mr. Airy in quartz itself (F, Plate VII.), when examined in convergent circularly-polarised light. We see that the quartz, considered as an ordinary uni-axial crystal, is able, owing to its peculiar and **totally distinct** effects upon plane-polarised light passing **through** it axially, *to show its own spirals*, which of course are double. **These are not** seen at all in parallel light ; **and on the** other hand, if we employ extremely convergent circularly-polarised light, they become as numerous **and** distinct as those of the calcite.

A crucial **test of this view of the case** readily suggests itself. **If it be well founded, we can** represent our quartz **artificially, as** it were ; since many fluids act similarly (§ 151) upon a plane-polarised **ray.** If therefore we take a column of such fluid of sufficient **length, and any ordinary uni-axial** crystal, the one will represent the axial properties, and **the other the ordinary doubly-refractive properties of** the **quartz ; and the** two ought **to give us** double spirals **; in** fact an adequate **column of fluid ought successfully to** replace the quartz **in all the foregoing experiments.**[1]

The rotatory effect of fluids is so inferior **to that of** quartz, that it is not easy to transmit sufficient **light to give good** projections through a column of **fluid of** adequate length. By employing a tube 8 inches long and 2 inches **in** diameter with plane glass ends, filled with oil of lemons (1 lb. of which costs about 10*s.* **6***d.***,** and is just sufficient to fill such a **tube) the object can however be effected.** We introduce **this next the polariser in lieu of the quartz.** In the crystal stage we **place the calcite or any other uni-axial crystal ;** and now introducing the quarter-wave plate between crystal and analyser, **we obtain at once the double spirals.** The fluid will also give the **same phenomena as the quartz with**

---

[1] It is probable that a **bar of heavy glass in the electro-magnetic field would give similar effects ;** but I have not as yet been able to **test the matter experimentally,** and there **is the** very interesting difference **between the behaviour of such** a bar and other **rotary** substances described **in §** 150. It seems scarcely probable **that this** difference would **affect the above** phenomena ; but the settlement **of** that point would be interest¹ng.

other crystals, its slightly yellow colour only slightly interfering with the effect, for the same reason that the figures fail to appear in homogeneous light. **Spirit of** turpentine is free from this defect, but requires **a** column of almost unmanageable length.

Finally, **it** may be mentioned that **Reusch's** artificial quartzes made of mica-films (§ 157) and Norremberg's artificial uni-axial crystals made of crossed micas (177), give in each case similar results to the natural crystals. So also does a circular disc of unannealed glass in parallel light.

CHAPTER XVII.

POLARISATION AND COLOUR OF THE SKY.—POLARISATION BY SMALL PARTICLES.

*Polarisation of the Sky—Light **Polarised by all** Small Particles—Blue Colour **similarly** Caused—Polarisation by Black Surfaces—Experimental **Demonstration** of the Phenomena—Multi-coloured Quartz **Images—Identity** of Heat, Light, **and Actinism**.*

185. **Polarisation of the Sky.**—On a clear day, in morning or afternoon, almost any of the colour phenomena we have now reviewed may be tolerably seen, by using the tourmaline close to the eye as analyser, and looking through the selenite or other object to the *sky* as polariser, in any direction at a tolerably wide angle with the direction of the sun, the maximum effect being at 90°. For instance, if the sun were due east, the greatest polarisation will be found anywhere in an arc extending due north and south. In the most favourable positions the quantity of polarised light is about one-fourth of the whole, and the rings in crystals can be seen very plainly with the sky as polariser. The direction of greatest polarisation of course depends upon the place of the sun; and upon this fact Sir Charles Wheatstone based the construction of a "polar clock," which gives the astronomical time by the effects upon slips of selenite in certain positions.

186. **Cause of the Phenomenon.**—Brücke **and** Professor Tyndall have beautifully explained not only this phenomenon, but also the blue colour of the sky, by proving experimentally that the light reflected laterally, or at right angles with the incident rays, from *any particles whatever sufficiently* **small**, **is** both polarised and of a blue colour. The blue **colour** is easily understood if we remember what we have always found, that the blue waves are the shortest or smallest. Hence, from particles **so** small as to be in commensurate relations with them, the smaller waves may be wholly reflected, while larger ones are broken up or shivered into fragments, **as it were,** and so destroyed; just as—to quote Dr. Tyndall's own image—pebbles on a shore reflect small ripples entire, while they scatter and break larger ones. A secondary proof of this is ready to hand in the colour of *transmitted* light. If it is partially robbed of its blue by these transverse reflections, the light transmitted ought to be more short of blue, or perceptibly yellowish, or **in** extreme cases reddish. That this is so we see every sunset, and also by the colour transmitted through any of **the** media presently mentioned.

187. **Polarisation by Small Particles.**—The polarisation **at an** angle of 90° with the incident ray, or at an angle of 45° with the surface **of** each **minute** spherical particle, has been considered a difficulty. **Sir** John Herschel remarked, that it supposes **an** index of refraction of unity; or that in the case of the sky **we** have to suppose reflection *in* air *upon* air. There will be no difficulty in supposing this, if we conceive the molecules of air reflecting light at all; and the angle of 45° is exactly what we shall expect, if we receive the reasoning advanced in § 120, or attach any weight to Sir David Brewster's arguments that the *real* angle of polarisation in all cases is 45°. We have only to suppose that in any case of this scattered reflection the

reflecting molecules are too small to exert any refractive influence, and the whole difficulty is solved. While some, therefore, have considered polarisation by small particles to be a *fourth* method of polarising light, it is not so considered here, but simply regarded as another case of polarisation by reflection.

188. **Black Surfaces.**—This seems also the only method of explaining the curious phenomenon of polarisation or analysation by a black surface, which was brought to my notice some time ago by Mr. Thomas S. Bazley. It is stated in many works upon physical optics that a black ribbon absorbs all the colours of the spectrum; but this is by no means practically the case in most instances with the bright colours of the solar spectrum, which have a peculiar and attractive effect—of course owing to scattered reflection—on a black ground. Further, however, if we pass the light from the lantern through a polarising Nicol and a plate of selenite, without any apparatus usually known as an analyser, and receive the light at right angles on a dead-black card, the colour due to the selenite will appear, though it will not if the card be white.[1] Hence the black card itself acts as an analyser; and we can only explain this on the supposition that the black colouring matter, by absorbing or quenching the reflections from the flat surface as such, allows us to perceive the comparatively feeble results of the light reflected from the small particles of carbon or other colouring matter. These particles are however so large, that it will be found the polarising angles considerably exceed 45°. And the colour is of course comparatively feeble.

189. **Experimental Demonstration.**—Professor Tyndall precipitates fine vapours[2] in an exhausted glass tube

---

[1] That is, with the card at right angles. An inclined white card analyses as a reflecting surface.

[2] Professor Tyndall has employed vapour from nitrite of butyl and

with glass ends; but simpler apparatus will amply suffice for us. Small particles in water show the same phenomena, and either (1) a *very little* soap, or (2) a few drops of milk, or (3) about six grains of resin in an ounce of alcohol, or (4) about five grains of pure mastic in the same, will answer very well. The mastic is best, and soap handiest for a sudden occasion; or a teaspoonful of the solution of coal-tar in alcohol known as Wright's Liquor Carbonis will give excellent effects if stirred into water. We may take a common glass lamp-chimney, 12 inches by 2 inches, grind one end flat and cement on it a flat glass plate, and fit a vulcanised stopper

FIG. 189.—Experimental Tube.

to the other. Fill carefully with filtered water, in which a very little soap is dissolved, or into which about a teaspoonful of the mastic in alcohol has been gradually poured while the water was violently stirred, either being *filtered* into the tube to remove dust, which mars the effect by reflecting common light. Mount the tube T in two semicircular notches of a cradle stand, as in C, Fig. 189, and adjust the tube horizontally in front of the plain optical objective—*i.e.* taking

hydrochloric acid, nitrite of amyl, bisulphide of carbon, and many other compounds. A friend of mine has obtained beautiful results from a whiff of tobacco smoke.

away the polariser—so as to throw the beam of light from the lantern along its axis.

If a dead-black board or sheet of card is held behind the tube, it is soon seen that it appears blue. The black background is not even necessary, for the tube *shines* with a sky-blue light, unless the quantity of solid matter is much too large. It will also be readily seen, on looking as nearly as possible at right angles towards the tube through a Nicol, that most of the scattered light is "polarised"; for rotating the Nicol in the hand alternately quenches and restores it, and a selenite held between the tube and the Nicol at the eye shows its colours. We have already learnt, however (§ 115), that any apparatus which will act as a polariser, may also be employed as an analyser; and by polarising the light first, and using the tube as analyser, we can make the phenomena visible to a number of people at once. Add the Nicol, N (Fig. 189), to the nozzle; the light from the lantern is now polarised, so that all who sit nearly at right angles with the tube can see the phenomena. As the Nicol is rotated, the light proceeding laterally from the tube is quenched or restored; and when *quenched* from a spectator on the same level, it is of course *brightest* to an eye looking down upon the tube from the top. Finally, if we hold a large quartz plate at Q, as the Nicol is rotated we get beautiful successions of colours in the tube.

190. **Multi-coloured Images.**—This is the simplest adaptation of the usual mode of performing this beautiful experiment; but there is a far better method—one not only easier, but which produces effects of surpassing brilliancy and beauty, and which, as a truly magnificent lecture demonstration, may fitly conclude this work.[1] Procure a

---

[1] I was originally indebted for this beautiful modification of the experiment to Mr. John Thomson, of Dundee, a very able demonstrator.

plain cylindrical glass jar on a foot, J (Fig. 190), 12 inches to 16 inches high, and 2 inches to 2½ inches diameter. Having cleaned it *bright*, filter the solution into that, and over it adjust the plane reflector, R, at an angle of 45°. The reflector throws the light from the Nicol, N, *down* through the fluid, which needs no glass plate, while the quartz can be laid on the top of the jar at Q. The first great advantage of this

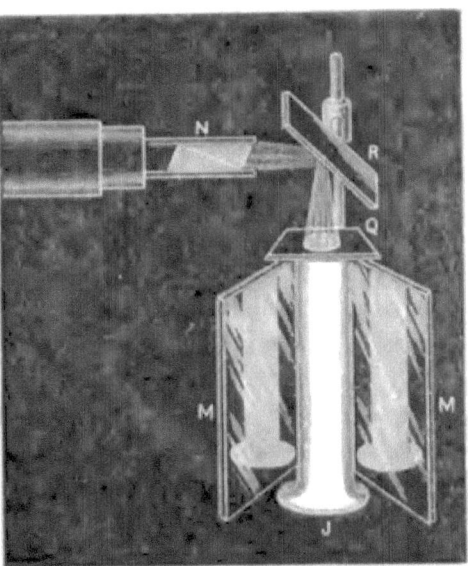

FIG. 190.—Multi-coloured Images.

method is, that the audience all over the room see the effects perfectly, if about the same *height* as the jar; whereas with the other and usual plan, only those who can look nearly at right angles towards the tube perceive much of the phenomena, which depend upon an angle of nearly 90°. Still further, however: if two additional silvered mirrors, M M,

The beauty of the effect is a great tribute to his ingenuity, but the method has hitherto been little known.

about the height of the jar by 6 inches or 7 inches wide, are arranged perpendicularly behind the tube, inclosing it as it were within a right angle, though not touching, they give, of course, by reflection, two additional images of the illuminated tube; but each of these, since the light leaves the tube from a different side, exhibits when the quartz is used a *different colour*, all three changing to successive colours with the rotation of the Nicol. With a large Nicol polariser embracing the full parallel beam from the lantern, the effect is finer still, and may be varied by employing a jar of rather greater diameter, and throwing the polarised light down through two apertures covered with quartzes of opposite rotations. In this case, in all but two positions of the analyser, there will be two beams of light in each image of the jar, glowing with different colours.

191. **Identity of Light, Heat, and Actinism.**—That the heat rays and chemical rays are subject to the same laws as luminous rays, as regards reflection, refraction, and dispersion, has been already stated (§ 88), and is a familiar truth proved in every camera every day. It only remains to state that they obey also the laws of polarisation and double refraction. If the two images which have passed through a double-image prism are tested with a thermopile, this is readily demonstrated, as is the fact that the ray is quenched whenever polariser and analyser are crossed. The actinic rays may be similarly tested with a sensitive plate, thus making the demonstration complete as regards all the rays of the visible and invisible spectrum, and proving that the sole difference between any of them is in period of vibration; some periods being more active in certain ways and some in others. Captain Abney has very recently shown that it is possible to obtain in a dark room a photographic image of a kettle heated far short of the luminous degree; or on the other hand, to impress a

sensitive plate with a photographic image of large portions of the spectrum through an apparently opaque plate of ebonite. And Professor Tyndall has carried the demonstration to the last degree of refinement, **by proving** experimentally that **a** plane-polarised beam **of dark heat,** filtered **of all** visible **rays by** a solution **of** iodine **in** bisulphide of **carbon,** is rotated, like the luminous rays (§ 150), by a powerful electric current, **or** when **the** glass **or other** diathermous material is placed in a magnetic field.

CHAPTER XVIII.

CONCLUSION.

THROUGH all the experiments now described, we have discovered that the phenomena and sensations we know as Light and Colour, when traced back and examined, found their ultimate explanation in forms of Motion. We were shut up very early to that conclusion: we were absolutely compelled to travel in our thoughts from what we "saw" to something we could not see at all, and to form mental images of invisible waves, whose undulations were propagated with incredible swiftness all around us. Later on we found phenomena which appeared to reveal to us the actual and precise *directions*, or orbits, of the vibrations in those waves; and applying to that hypothesis delicate and beautiful experimental tests, such as can be readily understood by any educated mechanic or other intelligent reader, we found our supposed orbits respond to those tests in every particular. The motions were, so far as we could judge from any possible mode of examination, modified, varied, resolved, or compounded, in all respects as our hypothesis led us to expect. This is the nature of the evidence, and we have thus reviewed in actual experiment the principal facts, on which is built up the Undulatory or Wave Theory of Light. The profoundest mathematical researches, applied to the most refined experiments varied in every possible way, have **so** far only confirmed that theory in every particular.

Let us fully grasp the grand conception; for there is no grander throughout the entire material Universe ! **All around us**—everywhere—space is traversed in all directions by myriads of **waves**. Not more surely does a nail take up from a hammer the force of a blow, than does each particle **of** Something take up and pass on the motion of the preceding particle. Heat, Light, Colour, Electricity—all alike **are** simply propagations of disturbance through that Something which we call Ether. Invisible themselves, these wonderful motions make all Things visible to us, and reveal to us such things as are. Take away from the diapason of these invisible waves those of any given period, and if **we** lose the dazzling whiteness which results from them all in due proportion, we but increase the soft splendour of the phenomena, **as** the hues of the rainbow appear before our eyes. Let them clash against, oppose, and so destroy one another; and even their very interferences, though dark shadows may cross our vision, produce amidst these forms and colours of almost unearthly beauty. Motion in the Ether accounts for all.

But we have taken another step from the seen to the unseen; for we have conceived and named this Ether. The name is of course nothing: but we cannot do without the thing itself—we *must* conceive it. **No** eye has seen it; no instruments can weigh it; **no vessel** can contain **it**; nothing can measure it; yet it must be there. "There?"—yea, here also, and everywhere. Absolutely invisible, it yet is the sole key to all physical phenomena; and the most recent, most widely received, and altogether most probable theory about Matter itself, **is** that *its* atoms are but Vortices in its infinite bosom. Ask for "absolute proof" of its verity, and there is absolutely none; and there are even about the conception itself some stupendous difficulties. The physicist has to endow his Ether with the most contradictory properties: he conceives it as rarer and more subtle than the most exhausted atmosphere, with the principal properties of a perfectly elastic fluid, and yet, withal, the chief distinguishing property of a solid! All these things do not deter him; and he believes implicitly in this Ether he has never seen and never will see, simply because without it he can explain no solitary phenomenon around him, while with it and its motions he can explain everything. Light is thus to him a Revealer of all Nature, both visible and invisible.

Another step further yet. The inquiry is irresistibly suggested, whether the comparison and the analogy may not go further, and afford us some revelation deeper still. That inquiry is strictly legitimate. If our universe be in truth an objective and conditioned manifestation of any absolute Source of all being, it should be thus; the Actual ought, in its limited measure, to reveal to us truly the Essential and Eternal. The student of nature, at all events, does hold expressly that if Nature has any Author, she must speak aright of Him if she speak at all; and as for the so-called religious man, while any book can only take a secondary place in such an inquiry as this, he also believes that it ought to be thus, since his book actually says so. The point of surpassing interest therefore is, whether as regards this question there is any definite agreement between these two, as to which Science can have anything to say, or possesses any means of judging.

What then do we find? We are bound here at least to ask the expounders of physical science first, for every reason. We inquire, therefore, what purely physical science, and experiment, and speculation—what *they* at present appear to teach us?

1. They tell us of an intangible, invisible Ether, which cannot be touched, or tasted, or contained, or measured, or

weighed, but yet *is* everywhere; which contains within itself the most essential properties of Matter, fluid and solid; and yet which is not matter, though it can communicate its own motions to matter, and receive motions from it.

2. They speak to **us** next, according to the **latest and** most widely received Vortex Theory of Sir William Thomson,[1] something vaguely about this Ether taking Form. They suggest to **us** how Vortices in **it may** appear to **us as** the atoms of Matter, which we do see, and feel, and handle; and which in this Form *can* be limited, and contained, and measured, and weighed; and **in** which **the** Ether may become, as it were, incarnate and embodied.

3. They tell us in the third place, of a mysterious Energy, which **also** takes protean forms, but which in one form or **other is** doing all the physical work of the Kosmos. Through it Ether acts upon Matter; and Matter re-acts upon Ether or upon **other** Matter.

And this **is** All; **and our** Light embodies them all and reveals them all. It *is* Motion, a form of Energy; it **is** Motion *in* the **Ether; and** it **is** invisible, inconceivable, unknown to us, *unless* Matter, to make **it** visible, be **in its** path. There are these Three and these only; each dis-**tinct** and separate; and yet the three making up One, a mysterious unity which cannot **be** dissolved.

So far the purely physical philosopher. Pondering **atten**tively this wonderful triune splendour which he has put before us, it may seem strange that he at least should sneer at *any* other Trinity in Unity, seeing the kindred mystery **in** which he himself acknowledges that he dwells. Ether: Matter: Energy: **no one** of these can be conceived of, hardly, apart **from** the others; **yet** each is separate and

---

[1] It **is** almost unnecessary to say that at present this is only a hypothesis. And equally so to remark that it makes more progress and receives further adherents every day.

distinct. Take away either, and what becomes of the Universe, as we know it or can conceive it? And yet this Universe at least is monistic—is one harmonious whole. The mystery of Nature is not only as great, but actually appears already to be of the *very same kind*, as that which theologians have taught us concerning the mystery of its Author.

For now we are at liberty to turn to the other, and ask him. He has known nothing of all this; never even dreamt of it, since it is the last growth of the nineteenth century. But, purely from an old Book he possesses, he too has, somehow or other, and long before the other, also gathered a conception, and even framed it into a set theological formula. It will be interesting at least to see what his conception is.

1. He tells us first, that he believes in an eternal, immortal, invisible, inconceivable, infinite Essence, the one Source and Father of all.

2. He believes that this first essential Being has in a mysterious way become embodied in a Second, in some inconceivable manner co-existent with and yet derived from Him; who is the brightness of His glory and the visible Image of His person,[1] and in whom and by whom all Things were made.

3. He affirms that these two work or act by and through a third, an equally mysterious Energy; whose operations assume many forms; who does all things, alike in matter and in spirit; who is as the wind, blowing where it listeth; and who finally brings all conscious agencies that yield to Him, into harmonious relation and equilibrium with all that surrounds them.

That is the creed of the Christian, however he came by it: more particularly indeed, it is the special creed of the

---

[1] "The very Image [*or* impress] of His substance."—*Revised Version*.

Trinitarian Christian, so much derided during **the** last twenty years. He also says **and** believes, like the other, that, although he cannot explain it, any more than the physical philosopher, these three are One. And strange to say, he too **goes so far** as to affirm that the **Motions** of the third originally produced that Light which we have found such a fascinating study; and that to him, also, that is an express symbol and revelation of the Three!

This is but a suggestion and inquiry, and **dogmatism is** not pretended from either side. But *if* there should be reality and fact behind the belief of both parties as we have listened to them, **is** there not indeed here **an** obvious, deep, fundamental, marvellous agreement? **More** than this: if there should **be true wisdom in what has** been taught us **by one of the most** popular teachers of modern philosophy;[1] **if it is true that** "Religion **and** Science **are** therefore necessary correlatives;" if it **is true** that "Force, as we know it, can be regarded **only as a** certain conditioned effect of **the** Unconditioned Cause—as **the relative** reality indicating to **us an** Absolute **Reality by** which **it is** immediately produced;" **if it is** further true that "objective Science can give no account of the world which we know **as** external, without regarding **its** changes **of form** as manifestations of something that continues constant under all forms;" and if it **is** finally **true as** regards Spirit and Matter, that "the one **is,** no less than **the** other, to be regarded **as** but a sign **of** the Unknown Reality which underlies both;"—if these conceptions of one whom all regard as at least a great thinker, embody anything more than a vague dream, **is** not this correspondence **we** have found, precisely **of the** sort we ought to have expected to find?

[1] **Mr.** Herbert Spencer. All the sentences quoted are from *First Principles*, 3rd edition; and the last one cited is the final sentence of all in that remarkable volume.

The comparison and the inquiry appear in any case to be singularly interesting. The student of Nature, at least, will not object to it; nor should he turn away repelled from the suggestion that Light may be to him a Revealer such as he has longed for, leading him into sight of, though not within, the inmost Secret of all. And as for the other, he too may perhaps learn to hear of Matter possessing "the promise and potency of every form of life" without resentment, and to attach to the phrase a new meaning, which may perchance be the basis of a great reconciliation that has been long and sorely needed. If what he believes be true, he will at least have learnt in another way that "the invisible things of Him since the creation of the world are clearly seen, even His eternal power and Godhead *being understood by the things which are made.*"

INDEX.

INDEX.

A

ABERRATION, spherical, **41, 62**; chromatic, 70, 83
Absorption, effects of, **140**; of colours, 112, 118; spectra, **129**; reciprocal with radiation, **135**, 146
Achromatic lenses, 83
Adams' arrangement for bi-axials, 318
Air, films of, 165
Airy on quartz waves, 315; quarter-wave plates, 297; spirals, 331
Analyser, 213; Delezenne's, **207**; size of, 244; glass, 254
Analysis, of waves, 101; of polarisation, 219, 227
Angles, of reflection, 27; of total reflection, 54; of polarisation, 222; of bi-axial crystals, 319
Anomalous dispersion, 81, 85, **107**; in crystals, 314, 320
Apophyllite rings, 314
Apparatus, 1-21; polarising, 242; for bi-axial crystals, 249, 316
Arragonite, conical refraction in, 324
Artificial quartz preparations, 303; uni-axial crystals, 313, 329; Nicol prisms, 257
Artists, mistakes of, 75, 121
Axes, optic, 226, 231, 325; relation of, 326, 338

B

BALMAIN'S paint, 144
Balsam and bezol, for mounting, **264**

Bands in spectrum of Newton's rings, 172; mica, **173**; selenite, 268; quartz, 286
Barton's buttons, 190
Becquerel on phosphorescence, 151
Bi-axial crystals, 231, 315; apparatus for exhibiting, 249, 316; theory of, 321
Biot's experiment on sonorous vibrations, 278
Bi-quartz, 286, 288
Bi-sulphide prisms, 16, 84
Black surfaces, polarisation by, 346
Blue and yellow, variety of effects from, 123
Brewster on the polarising angle, 223; his artificial crystals, 313
Brookite, peculiar dispersion of, **321**
Brücke on polarisation by small particles, 345
Bunsen's holder, **16**

C

CALCITE, 208
Calorescence, **150**
Camera obscura, 22
Care of calcite prisms, 252
Cascade, luminous, 55
Cauchy on dispersion, 108
Change in wave-length, 141
Chemistry, solar and stellar, 138
Chlorophyll, phenomena of, 114, 148
Chromatic aberration, 70
Circular double-refraction, 284, 292
Circular polarisation, 294; effects on crystal figures, 333

Clock, polar, 344
Collimator, use of, 84
Colour, 111; caused by suppression, 72, 112, 159; absorbed, reflected, and transmitted, 114; only a sensation, 118, 124
Colour-blindness, 127
Colours, not primary, 123; complementary, 116; of thin films, 158; of thick plates, 207; of polarised light, 260, 286
Combinations of mica and selenite, 328
Common light, theories of, 234
Complementary bands in spectrum of polarised light, 268, 286
Complementary colours, 116, 260, 286
Composite crystals, 326
Composition, of colours, 71; of vibrations, 295
Conical refraction, 324
Conservation of energy, 153
Continuous spectra, 131
Convergent rays in crystals, 309
Cord, displacement of, analogous to light, 239
Cornu's saccharometer, 289
Crossed crystals, 328; mica films, 329
Crossing films, effects of, 266
Crova's disc, 96
Crystallisations, 269; on the screen, 273
Crystals, uni-axial, 226, 230; bi-axial, 231, 315; rotatory, 290; composite, 326; irregular, 327; crossed, 328; rings and brushes in, 308; in circular light, 333; spirals in, 337
Crystal stage, 311
Cylindrical lens, 72

D

DE DOMINIS, experiment on the rainbow, 78
Deflection of rays, 57
Delezenne's analyser, 207

Descartes on the rainbow, 80
Designs for polarised light, 263
Deviation, minimum, in prism, 59
Diagrams for lantern, 18
Diffraction, 182; spectra, 191; in the microscope, 200
Direct vision prisms, 83
Direction of vibrations in polarisation, 224; in common light, 234
Disc, Crova's, 96; Newton's, 73
Dispersion, 66, 107
Dispersion, anomalous, 81, 85; in crystals, 314, 320
Dolbear's opheidoscope, 39
Doubled angle of reflection, 31
Double-image prisms, 209
Double refraction, 208; analysed, 227
Doubler, Norremberg's, 255; use in measuring films, 265, 298
Dove on common light, 237; on fluids, 292

E

ELASTICITY in the ether, 103; in crystals, 225, 321
Electro-magnetic rotation, 290
Elliptical polarisation, 236, 262, 297
Emission theory, 46, 91; test of, 169
Energy, conservation of, 153
Ether, the, 103, 353
Eye, easily deceived, 118; supposed mechanism of, 100

F

FILMS, colours of thin, 158; of mica, 264, 303, 329
Fizeau's experiment on velocity of light, 89
Fluids, rotation in, 287; producing spiral figures, 342; waves in, 155
Fluorescence, 145; and phosphorescence, 151
Forbes, Professor, on velocity of light, 90
Forces, result of two, 154

INDEX. 363

Foucault's experiment on velocity of light, 90; his prism, 251
Fox, on diffraction patterns, 185; on mica preparations, 264, 305
Fraunhofer's lines, 132, 136
Fresnel, on transverse vibrations, 217, 235; his theory of common light, 235; his mirrors, 178; his prism, **179**; his quartz prism, 284; his rhomb, 295; on quartz, 284; on rotatory polarisation in fluids, 292
Fuchsine, anomalous dispersion of, **85**

G

GASES, spectra **of, 130**; of incandescent, 133
Gas-burner in lantern, 4
Gas-regulators for lime-light, 9
Geological sections polarised, 273
Glass in sonorous vibration, 278
Glass piles, 215, 252
Goldleaf, colours of, **115**
Gratings, 183
Green, not a compound, **118**
Grey, nature of, 75
Gypsum, effects of heating, 321

H

HAMILTON on conical refraction, 324
Heat, effects of, 276, 322; identical with light, 142, 350; its vibrations visible, 33
Hemitrope crystals, 327
Herschel, on polarisation of the sky, 345; on reversal of phase in reflected light, 195; right and left handed crystals, 284; on a stretched cord, 241; on water-colour painting, 121
Hoffman's arrangement for bi-axials, 317
Huyghens' construction of waves, 101; experiment, 210; on double refraction, 217
Hyperbolic curves, 170, 328

I

ICELAND spar, 208
Identity of light, heat, and actinism, 142, 350
Images, 22; from mirrors, 29, 41; from lenses, 61; virtual, 28, 43, 62
Indices of refraction, 52, 230
Intensity, law of, 27
Interference, 154; conditions of, 158
Interference bands in spectrum, 172, 268, 286; in analysis, 282
Inverse squares, law of, 27
Inversion of images, 25, 42, **62**
Invisibility, of polished surfaces, 45; of light, 46
Invisible **rays,** 141
Iodide prism, 83
Irregular crystals, **327**
Irregular refraction, **181**
Isolating phenomena, need for, 24, 78, 173
Ivory balls as illustrating waves, 94

J

JAMIN'S prism, 257
Jellett's analyser, 289

K

KALEIDOPHONE, **37**
Kaleidoscope, 31
Kundt on sonorous vibrations, 281

L

LANTERN, 1; vertical attachment, 19; mounting, 12; diagrams, 18
Lantern polariscope, 4, 253
Lenses, 60; achromatic, 83; mounting, 14
Light, for optical experiments, 4; centering, 11; invisible, 46; a mode of motion, 91; theories of, 88, 91, 94
Lime-light, 6

Line spectra of gases, 133
Lines, Fraunhofer, 132, 136
Lines, thickened, 137
Lippich on common light, 238
Liquid waves, 155
Lissajous' experiment, 34
Lloyd on conical refraction, 324
Lockyer, on transmission of states, 93
Lommel on fluorescence, 148
Lycopodium, diffraction by, 186

M

MACH, on sonorous vibrations, 281
Magnesium light for fluorescence, 147
Magnetic rotation, 290
Matter, possibly vortices in the ether, 355
Measurement, of wave-lengths, 192; of molecules, 196
Mechanical illustrations, of undulatory theory, 108; of polarisation, 219
Metals, reflection from, 299
Mica, bands in spectrum of, 173, 268; designs in, 263; artificial uni-axial crystals of, 303, 329; work with films, 305; quarter-wave plates, 297; combinations with selenite, 328
Microscope, bi-axials in the, 318; diffraction in the, 200
Micro-slides, 250, 272
Mirrors, concave, 40; convex, 43; reflecting, 37; Fresnel's, 178
Mistakes of artists, 75, 121; of the eye, 118, 125
Mitscherlich's experiment, 226, 322; his saccharometer, 289
Mixed plates, 181
Mixtures of light and pigments, 118
Molecular constitution, 197, 291
Molecules, size of, 196
Mother-of-pearl, 188
Motion, light must be, 91; a form of, 352; a revealer of, 152
Mounts for lenses, &c., 13
Multiple images, 29, 59

N

NARROW slit, need of, 77
Natural colours residual, 115
Nature and her Author, 354
Newton's experiment on the spectrum, 65; in total reflection, 67; colour-disc, 74; rings 166; analysis of, 170
Nicol prisms, 243; care of, 252; artificial, 257
Nobert's gratings, 184, 197
Normal, the, in optics, 28
Norremberg's artificial mica-film uni-axial crystals, 329; doubler, 255; lenses for bi-axials, 317; mica and selenite combinations, 328

O

OBJECTIVE for optical lantern, 2
Oil-lamps, for lantern, 6
Opheidoscope, 39
Opposite rotations, 284
Optic axes, 226, 231, 325; relation of, 326, 338
Organic substances, 273
Oxide, colours in film of, 160
Oxygen, for lime-light, 7

P

P. ANGULATUM, effects of, 204
Particles, small, as polarisers, 345
Pendulum, experiments with, 295
Perforated cards, experiments with, 186
Persistence of vision, 73
Phase, reversal of, 195, 262
Phoneidoscope, 174
Phosphorescence, 144
Pillar-stands for apparatus, 13
Pigments, mixtures of, 118
Pincette, tourmaline, 311
Plane of polarisation, direction of vibrations in, 224
Plateau's soap solution, 162
Polar clock, 344

INDEX.

Polarisation, 212 ; analysis of, 219 ; **by** black surfaces, 346 ; of the **sky**, 224, 344 ; a test of the emission theory, 169
Polariscopes, 212, 245
Polarising angle, 222
Polarising apparatus, 242
Polished surfaces invisible, 45
Position, effects of, **in** prisms and lenses, 59, 71
Prazmowski's improved Nicol prism, 243
Press for glass, 274
Primary colour sensations, 122
Principal planes or sections, 229
Prismatic colours, experiments in compounding, 71
Prisms, 58; mounting, 15 ; bottle, 16 ; water, 59 ; effects of position of, 59, 70 ; direct vision, 83 ; double-image, 209 ; Foucault's, 251 ; Fresnel's 179 ; Jamin's, 257 ; Nicol's, 242, 252 ; Prazmowski's, 243
Propagation of waves, 101
Pulse made visible, 33

Q

QUARTER-WAVE plates, 297
Quartz, for experiments in fluorescence, 145 ; experiments in heating, 225 ; phenomena of, 283 ; spectrum analysis of, 286 ; artificial, 303 ; in circularly-polarised light, 306

R

RAINBOW, 78
Rayleigh, Lord, experiment on yellow light, 123
Rays of light, 22 ; compared with beams and pencils, 26
Red ink, experiments with, 115, 149
Reflecting mirror, 33
Reflection, law of, 27 ; multiple, 29 ; scattered, 44 ; total, 53 ; polarisation by, 215

Refraction, 49 ; law of, 51 ; index **of,** 52, 230 ; explanation of, 104 ; double, 208 ; irregular, 181 ; conical, 324 ; cylindrical, **324**
Religion and science, 357
Resolution of vibrations, 258
Result of two forces, 154
Retardation illustrated, 105
Retina, in vision, 100
Reusch's artificial quartzes, 303
Revealer, light a, 47, 130, 133. 139. 152, 200, 205, 278, 308, 354
Reversal of phase, 195, 262
Reversed lines, 134, 136
Reversibility of rays and phenomena, 28
Rhomb, **Fresnel's, 295**
Rings, Newton**'s, 166,** 263 ; in crystals, 308
Ripples shown **on screen, 38, 156**
Rocking spectrum, 76
Rotation, right and left, 284 ; in fluids, 287 ; electro-magnetic, 290 ; and molecular constitution, 291 ; of plane-polarised dark heat, 351
Rotational colours, of quartz, 283 ; fluids, 287 ; films, 300
Rotatory polarisation, 283 ; crystals, **290**

S

SACCHAROMETER, 287, 289
Salicine crystals, 270
Santonine, effects of, 127 ; as a polarising crystal, 272
Savart's bands, 331
Scattered reflection, **44**
Screens, 18
Sections of minerals, 273
Sel de seignette, peculiarities of, 321, 329
Selenite, colours of, 259 ; effects of heating, 321 ; two plates crossed, 329
Selenite designs, 263
Senarmont, on heating plates of quartz, 226 ; on compound salts, 329

Sensation of colour, **124**
Shadows, 26
Shells of waves in crystals, 230, 324
Sines, law of, 51
Single interferences **not** traceable, 156
Size **of molecules, 196**
Sky, polarisation of, 344
Slides for rolling balls, 95 ; for **exhibiting** wave-motion, 98
Slits, experiments with two, 183
Slits for spectrum, effect of width, **77**
Small particles as polarisers, **345**
Smoke in a jar, 47
Snell's law of sines, **79**
Soap films, 161, 198; and sound vibrations, 174
Sodium lines, 133, 135
Solar spectrum, 131, 136
Soleil's saccharometer, 289
Solutions of soap, **162**
Sonorous vibrations, **35, 174 ;** in polarised light, 278
Sound vibrations in soap films, **174**
Spectra, absorption, 129; continuous, **131** ; solar, **131** ; line, **133 ; reversed, 134;** diffracted **and prismatic, 191**
Spectrum, the, 65 ; a pure, 77 ; the invisible, 141
Spectrum analysis, 129 ; of Newton's rings, **170 ;** of soap film, **172 ;** of mica film, **173 ; of** selenite in **polarised light, 268, of** quartz, **286**
Sphinx, illusion of the, 45
Spirals, Airy's, 331 ; in a single quartz, **341 ; in crystals** generally, **337**
State of things, capable of transmission, 93
Stellar chemistry, 138
Stokes, on fluorescence, **145, 150 ;** on longitudinal vibrations, **240**
Strain, effects **of, 274**
Striated surfaces, **188**
Subjective colours, **126**
Sulphides, phosphorescence of, 144
Superposition films, 269, 303
Suppression the chief cause of colour, 72, 112, **159**

T

TABLE-STANDS, 16
Tank for refraction, 49
Taylor, Mr. Sedley, on **sound vibrations in films,** 174
Telescopes, reflecting, 43 ; refracting, 64
Ten-ion, **effects of, 274**
Thickened **lines, 137**
Thickness, of thin **films, 166,** 194 ; of mica-films, **265, 298**
Thick plates. colours of, 207
Thin films, **158 ; thickness** of, 166
Thomson's **experiment** with small particles, **348**
Three **spectra non-existent, 142**
Tidal **waves, 156**
Tisley's **phoneidoscope, 174**
Tonophant, 37
Total reflection, **53 ;** Newton's experiment in, **67**
Tourmalines, 213, 231
Tourmaline pincette, 311
Transmission of states, 93; of waves, 94
Transverse vibrations, 217
Trevelyan rocker, 33
Trinity, of nature, 354 ; of theology, 356
Tripod-stand for lantern, 12
Tuning-forks, experiments with, 34, 113
Turpentine, **experiments with,** 160, 181, **288, 343**
Tylor, **on mechanical** models of undulatory theory, 108
Tyndall, on calorescence, 150 ; refraction-tank, 49 ; subjective spectrum, **126 ;** fluorescence and phosphorescence, 151 ; analysis of Newton's rings, 172 ; polarisation **by small** particles, **345 ; rotation of** polarised dark heat-rays, **351 ; sonorous** vibrations, **278**

U

UNANNEALED glass, 278
Undulatory theory, 94 ; **see** *Wave*
Uni-axial crystals, 226, 230
Unseen, necessity of conceiving the, 353

V

VELOCITY of light, 88, 90
Vertical attachment to lantern, 19
Vibrations, absorbed, 140 ; direction of, in polarisation, 224 ; composition of, 295, 299 ; **resolution** of, 258
Virtual images, 28, 43, 62
Vision, supposed mechanism of, **100**

W

WATCH-GLASSES as concave mirrors, 43

Water-colours, effect **of,** 121
Water, colours in a film of, 165
Water prisms, 59 ; compound, 81
Wave-length, change in, 141
Wave-lengths, measuring, 192
Wave-motion, 94 ; slide for **show-** ing, 98 ; not in radial lines, **101**
Wave-shells in crystals, 230, 324
Wedges, of selenite, 263 ; of quartz, 288
Wheatstone's kaleidophone, **37** ; polar clock, 344
White light compound, 71, 116
Wild's sac barometer, 289

Y

YOUNG, on **mixed plates, 181** ; on diffraction, **182** ; on transverse vibrations, **217**

THE **END.**

LONDON:
R. CLAY, SONS, AND TAYLOR,
BREAD STREET HILL.

www.ingramcontent.com/pod-product-compliance
Lightning Source LLC
Chambersburg PA
CBHW051245300426
44114CB00011B/896